# *Lessons Learned*
## *A Guide to Improved Aircraft Design*

# Lessons Learned
## A Guide to Improved Aircraft Design

**Leland M. Nicolai**

**Ned Allen, Editor-in-Chief**
Lockheed Martin Corporation
Bethesda, Maryland

Published by
American Institute of Aeronautics and Astronautics, Inc.

American Institute of Aeronautics and Astronautics, Inc.
12700 Sunrise Valley Drive, Suite 200, Reston, VA 20191-5807

1  2  3  4  5

**Library of Congress Cataloging-in-Publication Data**
Record on file.

ISBN: 978-1-62410-381-0


Copyright © 2016 by Leland Nicolai. Published by the American Institute of Aeronautics and Astronautics, Inc., with permission. Printed in the United States of America. No part of this publication may be reproduced, distributed, or transmitted, in any form or by any means, or stored in a database or retrieval system, without the prior written permission of the publisher.

Data and information appearing in this book are for informational purposes only. AIAA is not responsible for any injury or damage resulting from use or reliance, nor does AIAA warrant that use or reliance will be free from privately owned rights.

To my children,
Jeffrey Stephan, Debra Leigh Nicolai-Moon, Noelle Michelle Nicolai-Ash,
Jeffrey Ray, and to Carolyn, my best friend and wife of 56 years.

I want to acknowledge the help and advice of some very smart and experienced people: Sherman Mullin (YF-22A Program Manager and retired president of Lockheed Advanced Development Company, The Skunk Works, Burbank, CA), Dr. G. Keith Richey (retired Chief Scientist, AF Wright Laboratory, Wright-Patterson AFB, Dayton, OH), Frank Cappuccio (retired Executive Vice President and General Manager of Lockheed Advanced Development Programs The Skunk Works, Palmdale, CA), Charles T. Burbage (F-22, X-35, and F-35 EVP/GM and retired former President Lockheed Martin Aeronautical Systems Company) and Eric Schrock (Deputy Director for Technology and Innovation, Lockheed Martin Aeronautics Co., Advanced Program Development, The Skunk Works, Palmdale, CA).

A special acknowledgement goes to my good friend
Grant Carichner (retired Program Manager, Lockheed Martin Aeronautics Co., Advanced Program Development, The Skunk Works, Palmdale, CA).
Grant provided advice, editing, and all graphics for the book.
He touched the book and made it a pleasure to read.

# A Note from the Editor-in-Chief

The creation of new free-flight cyber-physical systems, especially military aircraft, is perhaps the best documented, most thoroughly regulated, best understood engineering protocol in practice today. For such systems throughout their developmental evolution allow detailed analytical simulation, experimental verification (in wind and water tunnels), and subscale testing to verify performance. These are the archetypes of engineering excellence.

And still, in spite of such carefully crafted and controlled processes, serious mistakes are made, even in military platform projects, leading to poor and dangerous outcomes, abandoned and cancelled programs. Why and how could this ever happen? In this volume, Dr. Leland Nicolai, drawing on decades of his experience within real programs, spanned the full panoply of outcomes. He explores the subtle and profound reasons for why failures are so intractable in ten "Lessons Learned."

Dr. Nicolai's most recent experiences are rooted in his years as program manager, principle engineer, and engineering educator at Lockheed Martin's Skunk Works. His resume also includes years at other aerospace companies and at DARPA as well as teaching at the Air Force Academy, always working to refine and perfect the highest standards of engineering practice. He is among the U.S.'s premiere aeronautical engineering minds and authors, and for that reason, this well-crafted compendium of hard-earned program management wisdom has earned a place in the AIAA's Library of Flight.

The Library of Flight is part of the growing portfolio of information services from the American Institute of Aeronautics and Astronautics; it documents the crucial role of aerospace in enabling, facilitating, accelerating global commerce, communication, and defense. As demands on the world's aerospace space systems grows to support new capabilities, the series will seek to document the landmark events, emerging trends, and new views.

**Ned Allen**, Editor-in-Chief and AIAA Fellow

Chief Scientist, Corporate Engineering, Technology, and Operations
Lockheed Martin Corporation

Bethesda, Maryland

# FOREWORD

> Those who cannot remember the past are condemned to repeat it.
>
> *George Santayana, 1905*

Like many professions, aerospace engineers learn lessons many different ways. One of the most prevalent ways is by trial and error, which is certainly not the best approach. However, it has been around since the Wright brothers and, unfortunately, it will never go away. Most of us are not eager to set forth our lessons learned in writing, as it is often embarrassing. Most of us are employed by organizations not eager to publicize their lessons learned—they wish to have a more or less flawless public image.

We aerospace engineers are not noted for writing the history of our exploits. When we do it is usually the cleaned-up version of our project history. I know this to be true because like numerous others I have done it myself, which I admit here for the first time.

This book is an exception. It is a fine compendium of lessons from real-world aerospace programs. They show that knowledge of science and all of the relevant engineering disciplines are, as the mathematicians like to say, necessary but not sufficient. The reader can benefit significantly and avoid stumbling into the trial-and-error trap.

This is a practical book to read before you start on a new aerospace engineering project, and then re-read from time to time. It will be particularly useful to those who are just starting their aerospace engineering career and soon realize that many important issues are not resolved by equations, computer software, or other system design tools.

Now retired, after half a century in the aerospace business, I find the book interesting and intellectually stimulating. I believe you will also, and more importantly you will find it useful in your work and career.

**Sherman N. Mullin,** President (retired)
Lockheed Advanced Development Co.
The Skunk Works

# Contents

Preface . . . . . . . . . . . . . . . . . . . . . . . . . . . . . . . . . . . . . . . . . . . . . . . . . . . . . . . xiii

Chapter 1   Introduction – Going to Disneyland . . . . . . . . . . . . . . . . . . . . . . . . . 1

Chapter 2   Lesson Learned 1 – Display Ethical Behavior . . . . . . . . . . . . . . . . . 5

2.1   Background . . . . . . . . . . . . . . . . . . . . . . . . . . . . . . . . . . . . . . . . . . . . . . 5
2.2   Examples of Past Mistakes . . . . . . . . . . . . . . . . . . . . . . . . . . . . . . . . . . . 6

Chapter 3   Lesson Learned 2 – Have Fun . . . . . . . . . . . . . . . . . . . . . . . . . . . . 13

3.1   Background . . . . . . . . . . . . . . . . . . . . . . . . . . . . . . . . . . . . . . . . . . . . . 13
3.2   The Mistakes of the Past . . . . . . . . . . . . . . . . . . . . . . . . . . . . . . . . . . . 13

Chapter 4   Lesson Learned 3 – Listen to the Customer . . . . . . . . . . . . . . . . . 15

4.1   Background – Breaking the Code . . . . . . . . . . . . . . . . . . . . . . . . . . . . 15
4.2   Examples From The Past . . . . . . . . . . . . . . . . . . . . . . . . . . . . . . . . . . . 16

Chapter 5   Lesson Learned 4 – Challenge the Requirements . . . . . . . . . . . . 27

5.1   Background . . . . . . . . . . . . . . . . . . . . . . . . . . . . . . . . . . . . . . . . . . . . . 27
5.2   Examples of Flawed Requirements . . . . . . . . . . . . . . . . . . . . . . . . . . . 28

Chapter 6   Lesson Learned 5 – Populate the Design Team with Left and
            Right Brain Engineers  . . . . . . . . . . . . . . . . . . . . . . . . . . . . . . . . . . 39

6.1   Background . . . . . . . . . . . . . . . . . . . . . . . . . . . . . . . . . . . . . . . . . . . . . 39
6.2   Examples from the Past . . . . . . . . . . . . . . . . . . . . . . . . . . . . . . . . . . . . 39

| CHAPTER 7 | LESSON LEARNED 6 – THE INTEGRATED PRODUCT TEAM (IPT) WORKS – USE IT | 41 |

7.1 Background . . . . . . . . . . . . . . . . . . . . . . . . . . . . . . . . . . . . . . . . . . . . . . 41
7.2 Examples From The Past . . . . . . . . . . . . . . . . . . . . . . . . . . . . . . . . . . . 41

| CHAPTER 8 | LESSON LEARNED 7 – KISS (KEEP IT SIMPLE STUPID)…AS LONG AS YOU CAN | 43 |

8.1 Background . . . . . . . . . . . . . . . . . . . . . . . . . . . . . . . . . . . . . . . . . . . . . . 43
8.2 Examples From The Past . . . . . . . . . . . . . . . . . . . . . . . . . . . . . . . . . . . 43

| CHAPTER 9 | LESSON LEARNED 8 – WILLOUGHBY TEMPLATES WORK…USE THEM | 47 |

9.1 Background . . . . . . . . . . . . . . . . . . . . . . . . . . . . . . . . . . . . . . . . . . . . . . 47
9.2 Examples from the Past . . . . . . . . . . . . . . . . . . . . . . . . . . . . . . . . . . . . 49

**CHAPTER 10   LESSON LEARNED 9 – PLAN FOR VARIANTS . . . . . . . . . . . . . . . . . . . . . 51**

10.1 Background . . . . . . . . . . . . . . . . . . . . . . . . . . . . . . . . . . . . . . . . . . . . . . 51
10.2 Multiple Variant Designs . . . . . . . . . . . . . . . . . . . . . . . . . . . . . . . . . . . 52
10.3 Single-Variant Designs . . . . . . . . . . . . . . . . . . . . . . . . . . . . . . . . . . . . 53
10.4 AGM-86 ALCM and AGM-129 ACM . . . . . . . . . . . . . . . . . . . . . . . . . . 54

**CHAPTER 11   LESSON LEARNED 10 – EXHIBIT NYMPHOLEPSY . . . . . . . . . . . . . . . . . . 59**

11.1 Background . . . . . . . . . . . . . . . . . . . . . . . . . . . . . . . . . . . . . . . . . . . . . . 59
11.2 Examples of Nympholepsy . . . . . . . . . . . . . . . . . . . . . . . . . . . . . . . . . 59

**CHAPTER 12   EPILOGUE. . . . . . . . . . . . . . . . . . . . . . . . . . . . . . . . . . . . . . . . . . . . . . . . . 63**

**REFERENCES . . . . . . . . . . . . . . . . . . . . . . . . . . . . . . . . . . . . . . . . . . . . . . . . . . . . . . . . . 65**

# PREFACE

In our business of building airplanes for fun and profit there are two truths:

Truth 1—Every new aircraft program starts with a winning airplane design.
Truth 2—The "Best" design is not always the "Winning" design.

The customer gets inputs from his staff and ultimately selects the winning design. Each customer has a personal attribute that he wants in the design. I call this attribute the Measure of Merit—MoM. Determine this MoM and make it the centerpiece of your design, and you will have a winner.

The purpose of this book is to provide guidance to developing a winning airplane design. This guidance is in the form of Lessons Learned. A Lesson Learned has two sources:

1. It is sage advice based upon the experience from making (or observing) a mistake in the past.
2. It is the path forward of best practices from successful aircraft programs in the past.

History is our watering trough from which we drink, listen to others, and experience the mistakes and successes of past aircraft programs. It is important that past mistakes leave us with emotional scars because "He who forgets the mistakes of the past is destined to repeat them." Beware of the person who claims to have never made a mistake because he is DUE, and when it happens it will light up the sky.

I spent 23 years in the U.S. Air Force as an R&D officer where I wrote requirements for advanced aircraft, evaluated contractor designs in proposals, and was instrumental in the selection of contractor awards. I developed an aircraft for the Strategic Air Command while I was on joint duty with DARPA. In other words, I was the customer in the aircraft acquisition chain. I retired in January 1981 and entered the private sector.

I spent four years at Northrop Aircraft as an engineering manager, one year at Fairchild Republic, Farmingdale NY, as a vice president of engineering, and 29 years with Lockheed Martin at the Skunk Works as an engineering manager and Technical Fellow. Most of my industry experience was providing aircraft designs and their development to military customers. Aircraft that I have influenced are (chronologically) AGM-129A Advanced Cruise Missile, F-20A Tigershark, YF-22A Raptor, AGM-158A JASSM, and numerous black programs.

The bottom line is that having spent 57 years as both the customer and the supplier of advanced design aircraft, I have had ample time to drink from the watering trough as well as make and observe mistakes and successes. I have made good use of my time and feel qualified to offer up guidance for developing winning aircraft designs.

**Leland M. Nicolai**
June 2016

# Chapter 1

## INTRODUCTION – GOING TO DISNEYLAND

American athletes usually remark after winning the big one (Super Bowl, Olympic Gold, NCAA Basketball, etc.) "I'm going to Disneyland." The purpose of this book is to send the project team to Disneyland when their aircraft conceptual design wins the competition and moves into the next phase.

Our profession is one of designing and building aerospace products for fun and profit. This process can only start with a successful aircraft design. The design either beats the competition or you go home. A successful design starts with a project team that understands the requirements and time urgency, has credibility, and believes their design will be successful. It takes this kind of passion to generate the "fire in the belly" needed to work over the weekend doing one more trade study to find the next SR-71 (it is called nympholepsy and is discussed in Chapter 11).

This book is written for three audiences:

- First, the college aerospace engineering student doing a senior design or capstone course. Because most university professors have little industry experience, this book attempts to fill that void.
- Second, the industry program manager and practicing design engineer trying to produce a successful aircraft conceptual design for a proposal.
- Third, the government personnel involved in the acquisition process and charged with generating the requirements for a Request for Proposal (RFP).

Figure 1.1 shows the phases in a typical government program acquisition according to Department of Defense document DoD 5000.1 [1]. The time spans shown in the figure are optimistic because there are always gaps in the schedule during the process of government issuing an RFP, industry submitting a proposal, government evaluating the proposal and then selecting a winner. Commercial-only programs move much faster because the airplane builder controls the tempo and funding of the program. A typical time from the decision to build the aircraft (Milestone 1 or B for a government program or the start of preliminary design [PD] for the commercial program) to production is about 10 years for government and 5 years for commercial. Figure 1.1 shows the importance of the conceptual design phase–over 70%

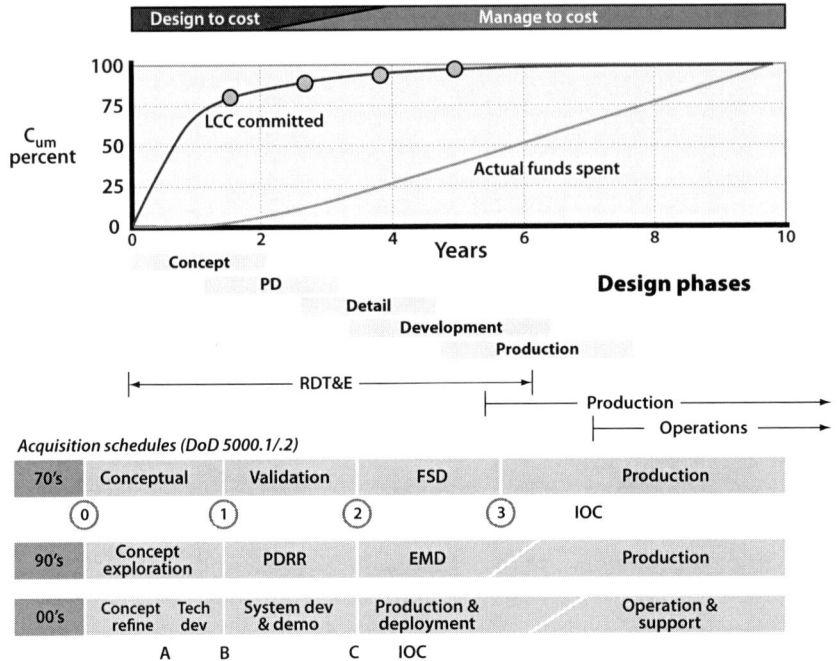

Fig. 1.1 Design phases integrated into the federal acquisition regulation DoD 5000.1/.2.

of the design features that drive life cycle cost (LCC) are selected during that phase.

The Lessons Learned discussed in this book are applicable to the conceptual design phase (either company funded or contract) up to design freeze at Milestone 1 or B. Most of the design work done during this period is analysis (mostly paper with some inexpensive wind tunnel and radar cross section [RCS] testing) by a team of about 20 engineers. Also during this period the company management and business development (BD) people interface with the government customer. If the Lessons Learned are implemented with zeal and passion, they will result in a world-class design. Following the Lessons Learned in this book will not guarantee success or that trip to Disneyland…but ignoring these lessons will most likely lead to a missed opportunity or repeating past mistakes. Most of the Lessons Learned are also good advice throughout the aircraft design and development phase.

The conceptual design phase determines the feasibility of meeting the requirements with a credible aircraft design. The conceptual design process is shown schematically in Fig. 1.2. The general size and configuration of the aircraft, the inboard profile and location of all the subsystems, and the general structural arrangement are determined during this phase.

# Introduction – Going to Disneyland

This book will focus on those Lessons Learned from previous aircraft conceptual design studies. A Lesson Learned has two sources:

1. It is sage advice based upon the experience of making a mistake in the past.
2. It is the path forward of best practices from past successful aircraft programs.

History is filled with mistakes made during the conduct of the conceptual design phase. Over his 56-year career in the aerospace profession, this author has been observer or a participant in some of the mistakes, leading to the 10 Lessons Learned presented in this book. The author also had the opportunity to watch project leaders skillfully apply the best practices to successful conceptual designs.

The importance of paying attention to the Lessons Learned in this book is

1. "He who forgets the mistakes of the past is destined to repeat them."
2. With best practices it is hard to screw up success.

So, the motivation is to learn from those that have gone before us and not repeat their mistakes but instead imitate their success.

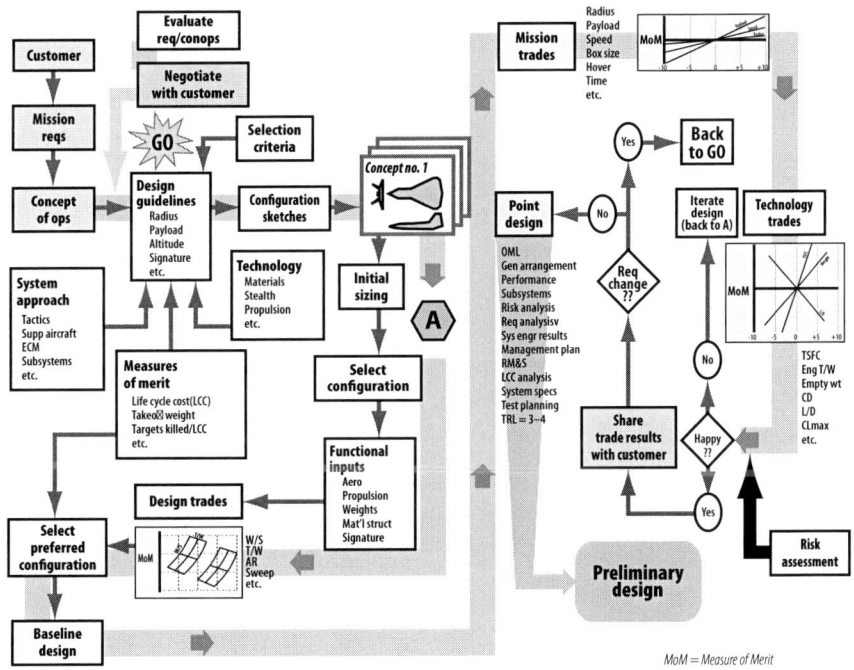

**Fig. 1.2  Aircraft conceptual design process.**

The book format will be to introduce the Lesson Learned, discuss its origin (either the mistake or the path of best practices), and give examples that led to the Lesson Learned. The first Lesson Learned is "Display Ethical Behavior" and rightly so as it trumps the remaining nine. The order of the subsequent Lessons Learned is chronological as they would appear in the normal conduct of an aircraft conceptual design study.

> From time to time throughout the book there will be a factoid that looks like this. The factoids are interesting comments outside the mainstream of the book story line that add depth and color to the message being presented.

# Chapter 2

## LESSON LEARNED 1 – DISPLAY ETHICAL BEHAVIOR

### *2.1 BACKGROUND*

Engineering is a proud and respected profession. It involves the design and fabrication of products for the benefit of mankind. This book focuses on aerospace engineering that produces aircraft and aircraft services. Because our products could malfunction and injure or kill someone, we have the responsibility to ensure that our analysis is correct, our design credible, our selection of materials matches the expected loading conditions, and the fabrication process meets the standards established by the U.S. Government Specification and Standard System, the Federal Aviation Authority, and other government agencies.

This responsibility is executed by engineers under a code of ethics that embraces the following fundamental canons:

1. Hold paramount the safety, health, and welfare of the public.
2. Perform services only in areas of their competence.
3. Issue public statements in an objective and truthful manner.
4. Act for each employer, customer, or client as faithful agents or trustees and avoid conflicts of interest.
5. Avoid deceptive acts such as withholding information or publishing dishonest status reports, financial documents, and technical/test reports.
6. Conduct themselves honorably, responsibly, ethically, and lawfully so as to promote the honor, reputation, and usefulness of the engineering profession and act with zero tolerance for bribery, fraud, and corruption.
7. Continue their professional development throughout their careers and provide opportunities for the professional development of younger engineers.

These canons are interpreted by the engineering companies and professional societies (IEEE, ASME SAE, AIAA, AIChE, ASCE, etc.) and issued to their members. The Accreditation Board for Engineering and Technology (ABET) publishes a Code of Ethics for Engineers to be used as guidelines in the accreditation process of engineering programs [2]. The most widely used interpretation is that published by the National Society of Professional Engineers (NSPE) [3].

Because of the unique activity of government civil servants in awarding and administering government contracts, the U.S. Government (and most states) have added the following to this list of fundamental canons.

8. Engineers in government service shall not use their position as stewards and keepers of industry proprietary, financial, and company sensitive data to give one company an advantage over another in competitive dealings or to be used for their own personal gain.

## 2.2 EXAMPLES OF PAST MISTAKES

Aeronautical history is filled with examples of unethical behavior leading to flawed products, unfair competitive processes, and companies going out of business.

**Ethical behavior is NUMBER ONE…it trumps the remaining nine.**

### 2.2.1 FAIRCHILD REPUBLIC COMPANY

The Republic Aviation Company was born in the late 1930s as Seversky Aircraft located in Farmingdale, NY on Long Island. It became Republic Aviation shortly before World War II and produced more than 26,000 military aircraft that fought in all major wars of the twentieth century, such as the P-47 Thunderbolt (WWII), F-84 Thunderjet (Korean war), F-105 Thunderchief (Vietnam war), and the A-10 Thunderbolt II. Republic was always a "one trick pony" operation having only one major program at a time. This made for great peaks and valleys in employment. In 1964 Republic delivered the last of 833 F-105s and the future looked bleak for the company. Fairchild Industries took over Republic in 1965 so that the company could survive.

In 1970, the Air Force issued an RFP for a low cost (less than $3M unit cost), easy to maintain, close-air-support aircraft. The new aircraft was to be designed around a 30-mm GAU-8 Gatling gun. The GAU-8 was as large as a Volkswagen "Bug" and was perfect for close-air-support. Northrop and Fairchild Republic won Milestone 1/B demonstration contracts to build two prototypes each of the YA-9 and YA-10 respectively. The Air Force selected Fairchild Republic to build 715 A-10As in 1973 and the future looked good for the company. In 1981 the production of the A-10 ended and the company looked around for its next program.

They didn't have to look far, because in 1981, the Air Force issued an RFP for the Next Generation Trainer (NGT). The NGT was to replace the Cessna T-37B used for undergraduate pilot training in the Air Force.

Fairchild Republic did their homework and prepared well for the NGT competition. They rolled up their cost to build two flight test aircraft and a static test article for the Development, Test, and Evaluation (DT&E) phase, and the cost was $132M. Competitive intelligence revealed that North American Rockwell was going to bid $120M.

So, Fairchild Republic dropped their price to $104M and won the fixed price DT&E contract in July 1982. Garrett Turbine Engines in Arizona was given a $115M contract to develop a high bypass turbine engine (F 109) for the trainer.

The T-46A team was in trouble right from the start as they tried to do $132M worth of work for $104M…and because it was a fixed price contract any cost over $104M would be borne by the company. To save money they got their suppliers to "bet on the come of a projected 650 unit production run" and invest their own money during DT&E. As a result, the suppliers had knowingly signed onto the program aware of the fact that they would be providing free parts for the first two aircraft along with associated testing and documentation that was also performed at their cost. This approach turned out to be flawed because, without financial incentives, many vendors gave the parts low priority, resulting in parts that were late and/or of low quality.

The T-46A team tried to cut corners to save money and ran into trouble, resulting in rework and more time and money. The late parts and rework caused the schedule to start slipping. When the Air Force and Fairchild Industries management asked how things were going, the FRC program team would withhold information, provide half-truths, and try to conceal the problems.

> The author was an "on the scene" observer of unethical behavior at the Fairchild Republic Company (FRC), Farmingdale, NY in the mid-1980s. At the center of this mistake was the development of the USAF Next Generation Trainer (NGT)—the T-46A. The mistake was the violation of fundamental canons 4 and 5 by FRC program management and engineers. The result was the destruction of people's careers and the demise of a proud aircraft company. An expanded reading of this ethical breakdown is presented as a case study in [4] pp. 833–841.

One of the milestones in the contract was a roll-out event. At roll-out, the first aircraft was to be ready for ground tests and system checkouts with all the equipment items installed. The roll-out was scheduled for February 1985, but at that date, the aircraft needed another eight months to be ready. The T-46A team chose not to inform the Air Force of the situation and proceeded with the roll-out.

The roll-out was a splendid affair with smoke, dancing lights, live music, and an invitee list that included all the bigwigs from Fairchild Industries and the Air Force, along with members of the New York State congressional delegation. The center of attention was test article 1 and from 50 feet the aircraft

Fig. 2.1 T-46A test article 1 at the February 1985 roll-out.

looked great, as shown in Fig. 2.1…but as you got up close it was apparent that all the equipment items were not installed and parts of the airplane were made of plastic, cardboard, and wood. The roll-out embarrassed Fairchild Industries and the Air Force. Slowly the full disclosure of the behind-schedule and over-budget status became known, which infuriated the Air Force.

Retribution came quickly. The Air Force advised the Secretary of Defense (SecDef) about the sorry state of the T-46A program and the deliberate withholding of important information by FRC program personnel. The SecDef ordered a contractor operations review (COR) for June 1985. The Air Force COR team was at Farmingdale from 4 June through 13 June and wrote up everything they saw wrong, from the level of company management down to safety violations in the cafeteria. Every single item that could be written up was in apparent retaliation for the botched roll-out. The result was the Air Force reducing the monthly progress payments from $6M to $3M and deferring the approval of the first production lot of T-46A aircraft. The reduction of the progress payments further aggravated the financial situation of the company. In addition, the company, already strapped for cash, had to fix the problems identified by the COR. The FRC program team continued to prepare the T-46A for first flight, but by this time they had blown past the $104M contract award and were spending their own money.

In September 1985 things went from bad to worse when the Secretary of Defense recommended canceling the T-46A program as part of a budget-trimming exercise. Shortly thereafter the management of Fairchild Industries announced that it was no longer interested in building airplanes and was looking for a buyer of the Farmingdale plant.

The first DT&E aircraft flew on 15 October 1985 (see Fig. 2.2) and the second about nine months later. Both aircraft conducted a normal flight test program at Edwards AFB, CA (500 h) and impressed those familiar with the T-37B with its performance and maintainability features. But it was clear to everyone that the T-46A was not going anywhere. The real winner of the flight test was the Garrett F 109-GA-100 turbofan engine, which performed great.

The end finally came on Friday the 13th in March 1987 when the Air Force formally terminated the T-46A program. The two DT&E aircraft are still at Edwards on display.

The sustaining engineering for the A-10A was transferred to Grumman in the fall of 1987. Fairchild Industries auctioned off the plant equipment and old tooling (most of it being sold for scrap) and closed the plant. Today the plant area is a movie complex and shopping center with very little indication that a major piece of U.S. military aviation history ever existed there. The decision by the T-46A program team to "low ball" the proposal price and to enter into a supplier "buy-in" subcontract could be viewed as poor program judgment and reconciled by the Air Force and Fairchild Industries management. But the

Fig. 2.2   T-46A first flight on 15 October 1985.

deliberate withholding of information that led to the embarrassing roll-out was inexcusable and an egregious violation of ethical behavior and mutual trust.

### 2.2.2 DARLEEN DRUYUN

Darleen Druyun was a career Air Force civil servant specializing in weapon system procurement. She joined the civil service ranks shortly after graduating from Chaminade University of Honolulu in 1972. She spent her career climbing the rungs of the Pentagon ladder where she cultivated an image as a hard-knuckled bargainer on billions of dollars in defense contracts with the nation's largest defense companies. She was called the "Dragon Lady" and took great delight in ridiculing corporate executives.

Darleen was nominated by President Bill Clinton in 1994 to be Principal Undersecretary of the Air Force for weapons procurement. Initially she was a good steward of industry proprietary, financial, and company sensitive data. She fostered a reputation as one of the defense contractors' toughest adversaries. But toward the end of the 1990s she formed a cozy relationship with Mike Sears, the chief financial officer of the industry giant Boeing Company [5].

In 2000 Druyun sent the resumes of her daughter, a recent college graduate, and her daughter's fiancé, a published PhD aeronautical engineer, to Boeing. They were both hired at elevated levels. Subsequently, in 2000, Darleen agreed to increase the size of the Boeing contract for C-17 transport planes by $412M. Two years later, she restructured the company's program to modernize 18 NATO airplanes used as airborne command posts and approved a $100M payment.

In 2001 the DoD was competing a contract to develop a sensor for U.S. weather satellites. A 2003 DoD audit revealed that Darleen had rigged the selection process and in four instances "manipulated complex proposal evaluation ratings to benefit Boeing and hinder the competition." In the same year Druyun picked Boeing over Lockheed to upgrade the avionics on the C-130 transport plane. The decision stunned industry analysts because Lockheed had built the airplane and was considered the most qualified to do the modernization.

In May 2003, the Air Force announced it would lease 100 KC-767 tanker aircraft to replace the oldest 167 of its KC-135 fleet. The 10-year lease would give the Air Force the option to purchase the aircraft at the end of the contract. In September 2003, responding to critics who argued that the lease was vastly more expensive than an outright purchase, the DoD announced a revised lease. In November 2003 the Air Force decided it would lease 20 KC-767 aircraft and buy 80 tankers.

Buying one KC-767 outright cost $150M. The contract called for 100 aircraft being purchased or leased at an average price of $370M per plane. Therefore, if the contract had been executed, the Air Force would have paid

Boeing much more money per aircraft than it would have had to if the aircraft were purchased individually.

In November 2003 Darleen retired from government service and accepted a vice president position with Boeing paying $250,000 per year. A month later the Pentagon put the tanker replacement program on hold while it investigated allegations of corruption and conflict of interest by Druyun. In the ensuing investigation Darleen pleaded guilty to inflating the price of the tanker contract, passing Airbus competition sensitive information to Boeing, rigging the selection criteria on several billion dollar competitions to favor Boeing, and in general showing favoritism toward Boeing.

In the end Darleen Druyun was fired from Boeing and went to jail for nine months; Mike Sears was found guilty of complicity, fired from Boeing, and went to jail for four months; Boeing was fined $615M; and numerous Boeing contracts were voided and recompeted [6,7].

# Chapter 3

## **LESSON LEARNED 2 – HAVE FUN**

### *3.1 BACKGROUND*

If you don't like airplanes…do something else. If what you are doing is not fun…walk away because life is too short to spend 8 hours a day doing something you don't like. What you are doing must not be a job…it must be a passion:

If you enjoy what you are doing…it is a win-win for you and your company.
You think about the problems at work during your commute and on weekends.
You put in that extra effort to make sure your work is correct.
You are proud of your work and want to tell people about it (check the classification first).
You can overlook the frailties of the people you work with.
You can almost understand the seemingly clueless decisions made by the "technically lightweight" management…such as their preoccupation with process, their aversion to taking a risk, and their focus on the near term.

### *3.2 THE MISTAKES OF THE PAST*

The author in his 56 years in aerospace saw countless bright, energetic, and innovative young engineers leave the ranks because they did not have interesting and challenging work. They were not having fun and so they left their companies and became astronauts, directors, vice presidents, and other persons of influence. It is said that "people make the difference." These young people were denied the opportunity to enhance the value of the business unit. They could have turned a mediocre business unit into a dynamite organization.

The mistake that the aerospace companies made was to let their young engineers (less than five years with the company) leave the company. They forgot why the young people became engineers in the first place. Young people become engineers to be challenged intellectually, to develop professionally, and to build something…in our case pieces and parts of an airplane. The bottom line is not to get rich, but rather build or be around airplanes, live comfortably, and have fun.

Most companies have an engineering leadership development program (ELDP), which rotates promising young engineers through several 6–12-month assignments in different disciplines. The ELDP is a good program, but less than 20% of the young engineers get selected for the program. The aerospace companies also have training and travel budget for young engineers to develop professionally, but here again, less than 50% of the engineers get to take training classes and/or attend technical conferences. The reduction of the new business funds to sales ratio from 3% (the aerospace standard for decades) to the current levels of 1.5% not only limits the retention of the young engineering talent, but it sends the wrong message. The message that the companies need to send to their young engineers is, "Yes, we care about your professional development and here is what we are doing about it."

Another mistake companies make is not letting the engineers build the product that they designed. There comes a point in all projects when the decision is made to make or buy the product (the dreaded make/buy decision). Too often the decision is to buy the fabrication of the product based upon cost. What happens then is

1. The fabrication of the product pieces is outsourced to a supplier that had nothing to do with the design and fabrication process. The fabrication is a "build-to-print."
2. The young engineer is denied the real-world experience of working with the shop personnel and seeing his design come alive and learning what it takes to make a manufacturing friendly, low-cost product design.
3. After several years of outsourcing the fabrication work, the company loses the capability of building things in-house.

Every employer of engineers needs to know at all times how the engineers feel about their work…are they having fun, are they challenged intellectually, or is it just an eight-to-five job. Also, know what is the message being received by the engineering community from the management.

It is incumbent upon any company or government agency to insist that their young engineers have a mentor or confidant of their choosing (not in the management chain) that they can go to for career advice. The mentors would meet at least once a month with their young engineer for a status report and general "how's it going?" check-in. The meetings would discuss career goals and plans for professional development. The results of the meetings would be confidential, but anonymous concerns and issues would be passed up the management chain to provide a "gut check" on the message that management is sending to their troops. The whole purpose of the mentoring program would be to help the young engineer find exciting and challenging work, develop professionally, have fun, and stop the flight of the best and brightest from the company.

# Chapter 4

## LESSON LEARNED 3 – LISTEN TO THE CUSTOMER

### 4.1 BACKGROUND – BREAKING THE CODE

We build airplanes for fun and profit. The profit motive implies that somewhere there is a customer who will buy our product if it meets his requirements. But our competition will most certainly meet the customer's requirements as well. So, how do we ensure that the customer will select our airplane?

#### 4.1.1 REQUIREMENTS AND MEASURE OF MERIT (MoM)

Requirements are quantified aircraft characteristics that describe the aircraft's required performance, cost, RM&S (reliability, maintainability, and supportability), etc. Typical requirements are range = 800 nm, payload = 5000 lb, unit cost = $35M and maintenance index = 15 man hours/flight hour. The requirements are generated by a customer committee and published along with the concept of operations (ConOps) and system architecture.

The measure of merit (MoM) is a qualitative aircraft characteristic that describes what is really important to the customer. The MoM might be one of the requirements in a published list, but it is not identified as the number 1 priority. Typical MoMs are low life cycle cost, high survivability, high maneuverability, minimum take-off weight, etc. The MoMs are important to the customer, but he is not able to quantify them. It goes without saying that aircraft appearance is a MoM because no one wants an ugly airplane. The MoM is a tie breaker when all of the proposals meet the requirements.

A question arises that if the MoM is so important…why is it not published? The answer is that the MoM is a very emotional thing with the customer and publishing the MoM would mean getting committee agreement which is usually not possible. Also, the customer can change during the acquisition and the new "customer" may not agree with the MoM. Because MoMs are not published, it makes this Lesson Learned extremely important, as listening to the customer is the only way you will learn of his MoMs.

Some Lessons Learned have rules imbedded in them. While Lessons Learned are good guidance and optional…rules are mandatory.

**Rule 1 – Meet the requirements…But Design for the MoM!**

If you guess right on the MoM you win…if you guess wrong you lose. The situation is made even more difficult by the very real possibility that the customer can change during the competition and you find your design chasing the wrong MoM. The message takeaway from this is to keep listening to the "right" customer throughout the acquisition and stay loose in your design.

#### 4.1.2 Who is the Customer?

The customer could be anybody…but we have to get it right because the customer owns the MoM and will make the decision about who wins the competition. In the commercial sector, it is pretty clear who the customer is. In the commercial transport or business jet market, it is the airline management (CEO and Board of Directors) with input from the operators. In the general aviation (GA) market, the customer is John Q. Public.

In the government sector (military), the ultimate decision maker is the Secretary of Defense (SecDef). The SecDef delegates responsibility to the civilian assistant secretary for acquisition in each military organization who, in turn, delegates to a program executive officer (PEO). The PEO is the source selection authority (SSA) for the acquisition. From here on, it can get very confusing as to whom the SSA is listening to when the requirements are developed and the final selection is made. It is critical to decide who the SSA is listening to because that person is the elusive "customer" that owns the MoM and needs to be listened to. Usually that customer person is the lead technical evaluator during the proposal process. It really gets confusing when several different people think they are the customer and like to hold court with the competitors and pontificate. Sometimes during the competition there is a change in the administration (such as a new SecDef), or there is a management realignment (perhaps a new PEO), or there is political maneuvering inside the program, or the world situation takes a turn…and we have to start over with a new customer and MoM (the requirements will most likely not change). Determining the "right" customer is absolutely critical.

It is critical to understand the requirements and the MoM as they drive the design. It is important to listen carefully in order to understand the requirements and hopefully help shape the requirements and conops. The MoM will come from the customer which could be the buyer (contracting officer), the cost community (GAO), the user (which could be multiple communities such as air superiority, ISR, strike, electronic warfare, etc.), or the maintainer (logistics).

### 4.2 Examples From The Past

#### 4.2.1 The USAF Light Weight Fighter Program (1972) – A Lockheed Mistake

In 1972 the U.S. Air Force launched a program for a light weight (20,000-lb weight class), low cost, daytime, air superiority fighter to complement the F-15 Eagle, providing the USAF with a high/low mix option. The aircraft was to have a radius of 250 nm, good turn rate, and acceleration, and be optimized for air combat at speeds of Mach 0.6-1.6 at altitudes of 30,000–40,000 feet. The payload was two AIM-9 Sidewinder IR missiles and an M-61 20-mm cannon with 560 rounds (total payload including a single pilot was 1500 lb). The program was called the Light Weight Fighter (LWF) program. It was not clear whether the customer was the cost community (GAO), Congress, the fighter community, or somebody else.

Industry was invited to present their LWF concepts to the Air Staff management before RFP submittal. Clarence "Kelly" Johnson, leader of the Lockheed Skunk Works, presented the Lockheed 1600 Lancer shown in Fig. 4.1. The Lancer was an F-104 with a large wing to give it a lower wing loading and outstanding turn capability. The Lancer had a top speed in excess of Mach 2.0 with all-weather capability and weighed more than 35,000 lb. It was more than the Air Force wanted and threatened the F-15. The Air Force leadership refused to consider Kelly's proposal. A high-ranking general officer remarked, "That #@&$# Kelly.... He never listens. He thinks he knows what we want better than we do." It was clear that at that point in time the customer was the Air Staff and the MoM was "preserve the F-15." Shortly thereafter, the customer changed to the fighter pilot community and the MoM became world-class maneuver capability at low cost.

**Fig. 4.1 The Lockheed CL-1600 Lancer.**

### 4.2.2 LONG-ENDURANCE MULTI-INTELLIGENCE VEHICLE (LEMV) – A SKUNK WORKS MISTAKE

In 2009 the U.S. Army issued an RFP for a long-endurance, theater ISR platform called LEMV. The requirements for the platform were an endurance of 21 days at 20,000 ft collecting battle field intelligence from 2500 lb of the most advanced cameras, sensors, and spy technology. Because of the requirement for 21 days TOS (time-on-station), the platform solution needed to be a helium-filled airship.

The Lockheed Martin Skunk Works (LMSW) was experienced in the ISR mission area and in the design and integration of airships. The Army indicated that the focus of the program was the rapid demonstration (in 18 months) of the 21-day endurance of the airship platform. LMSW started early, talking and listening to the Army customer in charge of the airship. They even helped the customer shape the requirements and concept of operations for the airship. As the end of 2009 drew near, LMSW was very well prepared for the RFP and even had the conceptual design of the airship completed. Their offering included an existing ground control station (GCS) providing the lowest cost and risk and shortest time to field a solution in 18 months.

Several months before the RFP was released, the Army PEO/SSA (Program Executive Officer/Source Selection Authority) began including the Army UAV program leadership in his source selection team. The Army operates thousands of UAVs and one of the critical objectives for the UAV team was the development of a common GCS that would operate all Army UAVs. While the LEMV program was initially started with the priority to rapidly field a long-endurance ISR platform, the unmanned nature of the airship led the program to include the UAV members who had different priorities (MoMs). The LEMV RFP was published with minimal changes to the original requirements and conops, but it was a major shift in the PEO/SSA MoM from a rapidly deployed long-endurance ISR platform to the development of a common GCS.

The PEO/SSA selected the Northrop Grumman Corporation (NGC) who had exclusive rights to the first-generation common GCS already in use by the Army. NGC proposed to upgrade the GCS as a central feature of their proposal. As it turned out, NGC had been talking with the Army UAV team and had helped them shape the program to emphasize the ground control/sensor fusion element of the program. NGC had poured most of their bid and proposal resources into defining the GCS and subcontracted with a small British company to build the airship shown in Fig. 4.2.

> Lockheed Martin seriously considered protesting the LEMV award. They decided not to protest because of the ill will it would create with the Army and the relatively small value of the contract.

## LESSON LEARNED 3 – LISTEN TO THE CUSTOMER

Fig. 4.2   Rollout of LEMV hybrid airship.

In the end, the program was cancelled in 2013 due to the British company overrunning the program's cost and schedule. In addition, the airship was 12,000-lb overweight, which reduced the endurance from 21 days to 4. However, the Army did get an upgraded common GCS for their UAV operation.

The takeaway from this historical example is to make sure that you pay close attention to the political maneuvering inside the program and are listening to the "right" customer.

#### 4.2.3 THE BOEING 777 JET TRANSPORT (1986) – A SUCCESS

In the commercial airplane industry, the airplane builder would typically do a market analysis to determine what the airline industry wanted. The market analysts in the Boeing Commercial Airplane Company in Seattle were masters when it came to determining the requirements for the next transport aircraft. They would set the requirements, the engineers would design the aircraft, the International Aerospace Workers (IAW) would build the airplane, and the airlines would "take it or leave it." The Boeing Seven Series of large transport aircraft (707, 727, 737, 747, 757, and 767) was very successful and gave the Boeing Company the major market share of transport aircraft well into the 1980s.

In 1986 the airlines were telling the transport industry that they wanted an aircraft that was bigger than the 767 but smaller than the 747. Boeing started planning for a new transport called the B-777 or "Triple Seven." The B-777 would be a twin-engine, long-range (at least 5500 nm), 300+ passengers, twenty-first-century airplane with a 3-h ETOPS (extended twin-engine operations).

The FAA, European Aviation Agency, and Japanese Aviation Agency ETOPS rule is that the twin-engine airplane is limited to operating over routes that contain an adequate airport that can be reached in X hours with one engine operation. The CONUS to Hawaii routes required X to be at least 2 h with 3 h preferred.

In the 1990s Boeing and Airbus would dominate the large jet transport market. The Triple 7 would be in a head-to-head competition with the Airbus 340 in what was called the "airliner war." In 1989 Boeing made the decision to spend $5B and five years developing the Triple 7. This represented a substantial commitment and risk for the company. The Triple 7 program had to succeed or Boeing risked losing its leadership position in the commercial transport market.

Boeing kicked off the B-777 program with a most unusual strategy. For the first time in Boeing Commercial Airplane history, eight major airlines (All Nippon Airways, American, British Airways, Cathay Pacific, Delta, Japan Airlines, Qantas, and United Airlines) had a significant role in the development of the airplane. This was a major departure from historical practice for Boeing who, in the past, would conduct the design process with little input from the customer. The eight airlines came to be known as the "gang of eight" and Boeing called the program management style "working together." The consensus of the "gang of eight" was that they wanted a cabin cross-section close to the B-747s, a capacity of at least 325 passengers, fly-by-wire controls, a glass cockpit, flexible interior, and 10% better seat-mile costs than the Airbus 330/340.

The Boeing management, engineers, and production workers employed "best practices" along with "working together" to roll out a world-class transport. The first flight was in June 1994 and the aircraft was certified by the FAA and JAA in April 1995 with a 3-h ETOPS (first time in aviation history…this meant that the first customer, United Airlines, could use the B-777 immediately on its CONUS to Hawaii routes).

In the decade between 1999 and 2008, Boeing delivered 956 Triple 7s to Airbus' 375 A340s. The Triple 7 has replaced the B-747 as the most profitable commercial jet in the Boeing stable. In November 2005 the B-777-200LR shown in Fig. 4.3 flew 11,664 nm on a special flight from Hong Kong to London, setting a world record for the longest nonstop flight for a commercial jetliner. In December 2011 the FAA certified the Boeing 777 airliner for a 5.5-h ETOPS…the longest for any current two-engine jet transport. For an expanded discussion of the Boeing 777, see [4], pages 603–612.

Fig. 4.3   B-777-200LR.

#### 4.2.4 ADVANCED TACTICAL FIGHTER (1986) – A LOCKHEED MARTIN AERONAUTICS SUCCESS

In February 1986 the U.S. fighter aircraft industry responded to an Air Force Milestone B RFP for a new air-to-air fighter to replace the F-15. The RFP requirements called for "super maneuver" and "super cruise" (the ability of the aircraft to cruise at $M > 1.2$ on dry power) and "super stealth" (true twenty-first century signature levels). In addition the Air Force wanted a TOGW < 50,000 lb, unit fly-away cost of $35M (1985 dollars), and a runway length of 2000 feet (later changed to 3000 feet). The Lockheed Skunk Works entry was a very low observable configuration with a trapezoidal wing and canted vertical tails, as shown in Fig. 4.4.

Fig. 4.4   Lockheed ATF entry in 1986.

The Lockheed Skunk Works and Northrop Aircraft Division won Dem/Val (Demonstration/Validation) contracts to build two prototypes each and have a fly-off.

At the beginning of Dem/Val in November 1986, Lockheed teamed with Boeing and General Dynamics/Fort Worth (builder of the F-16 Fighting Falcon). Northrop, in turn, teamed with McDonnell Aircraft (builder of the F-15 Eagle). The two teams appeared evenly matched, with each having strong fighter and low observable experience.

As the configuration design trades were being conducted for the YF-22, Lockheed observed that the requirements for super cruise and super stealth were in conflict with those for super maneuver. Super cruise and super stealth wanted small surface areas (low wetted area for low supersonic drag and traveling wave RCS), but super maneuver wanted a big wing (low wing loading for low $C_L$ during maneuver and low induced drag) and big horizontal and vertical tails (good pitch and directional control authority) giving a high wetted area. In discussions with the USAF it became clear that the customer was the fighter pilot community and they wanted a fighter pilot's airplane first and foremost with as much super cruise and super stealth as they could get without compromising a world class maneuver capability. Between July and October 1987 the Lockheed team changed their Milestone B winning design to a configuration having a large-clipped diamond wing and horizontal and canted verticals, as shown in Fig. 4.5 [8]. The Lockheed team made "maneuver with reckless abandon" their MoM. This MoM actually compromised the super cruise Mach number and RCS, but still met the requirements. The addition of pitch thrust vectoring (not required) added weight and cost to the YF-22 but gave the airplane high angle-of-attack (AoA) maneuvering that was unprecedented.

The high AoA capability (shown in Fig. 4.6) allows the pilot the freedom to point the nose almost anywhere and not worry about the aircraft departing controlled flight (maneuver with reckless abandon). Pointing during air combat maneuvering (ACM) would be establishing a cut-off angle to set up for a

Fig. 4.5  Comparison of the Lockheed YF-22 (l) and the Northrop YF-23 (r).

## LESSON LEARNED 3 – LISTEN TO THE CUSTOMER

Fig. 4.6   High AOA (60-deg) operation of YF-22.

gun kill or an IR shot (if only for a fleeting moment) or to simply point at the opponent to threaten him into making some kind of tactical blunder. The procedure would be to pitch the F-22 to 60-deg AoA, roll around the aircraft axis to line up with the threat, and then slam the nose down with the pitch thrust vectoring. The significant technology here is the combination of departure-free 60-deg AoA and pitch thrust vectoring. There have been other aircraft capable of controlled flight at high AoA for decades (i.e., the Soviet Su-27 and Su-37 and the Northrop F-20 Tigershark), but their aerodynamic pitch recovery was so poor that it took many seconds to get the aircraft nose back down, giving the opponent time to get away.

Each design team defined their own flight test. The Lockheed team chose to demonstrate the high AOA capability of the YF-22 and launched the AIM-120 and AIM-9 missiles from the missile bays. The Northrop team did neither. The YF-23 (shown in Fig. 4.5) was a beautiful airplane and actually beat the YF-22 in super cruise Mach and RCS, but the fighter pilots preferred the maneuverability of the YF-22. The F-22A is now operational with the USAF.

### 4.2.5 POLITICS IN THE PROCESS

The government, by definition, is a very political organization. So, it stands to reason that the DoD and the military services would be buffeted by political pressures. One can only wonder what political maneuvering takes place during a major weapon system acquisition. The example in Chapter 2 of the ethical breakdown of Darleen Druyun is clear evidence that personal agendas run rampant at the highest levels of the government. The change in program focus from the airship platform to the common GCS in the LEMV program (Section 4.2.2) was probably politically driven. It is well known that the industry lobbyists have considerable influence on the federal acquisition personnel.

**So, what is the lesson to be learned by industry from this observation?**

Industry needs to trust in the federal acquisition system and its regulations…otherwise there is no system. Industry also needs to trust in the federal acquisition process and the government personnel that are executing it. Industry needs to assume that the government people will be good stewards of their position and will abide by canon 8 in Chapter 2 (Section 2.1). But industry needs to be an informed and astute participant in the acquisition process. Industry needs to be very familiar with the customer at all levels (civilian deputy, PEO, customer communities) and the item being procured. A potential check list follows:

- Who are the potential customers (i.e., those making the selection decision) in this procurement?
- What are the different customer's job histories?
- How experienced are the customers with the relevant mission area?
- Did the customers come to the government from industry, military, or directly from college?
- What/who is the incumbent for the item being procured?
- What is the alternative(s) to the item being procured?
- What do the customers think about your company? About the other companies?
- Do the customers have a policy of spreading the work out across the industry base (i.e., is it your turn)?

The message takeaway for this chapter is to determine who the "right" customer is and then listen to him…but pay attention to other potential customers. Also, even after you have won a MS B contract, keep listening to the customer and be ready for a heading change.

---

**CUSTOMER DECISIONS ARE NOT ALWAYS BASED ON TECHNICAL EVIDENCE**

The GBU-15 is a 3650-lb unpowered glide precision guided munition (PGM) used to destroy high-value targets. The PGM uses an existing 2000-lb munition, either the Mk 84 general purpose (GP) bomb or the BLU-109 penetrating warhead. The weapon uses EO/TV or imaging IR and a data link. An operator (either airborne or on the ground) can guide the GBU-15 using the EO or IR imagery to the target or acquire the target and lock the guidance and the weapon automatically guides to the target. The weapon was developed by Rockwell International in 1974.

The GBU-15 came up for its MS 3 production decision (see Fig. 1.1) in 1976. The decision to produce a weapon system depends upon the results of an Analysis of Alternatives (AoA). The AoA determines if there is an alternative to the weapon system being considered for production. Typically the alternative is a performance issue....Does there exist an alternative having equal performance and effectiveness at a lower cost?

I was a lieutenant colonel on the Air Staff, Studies & Analysis in 1976. My boss was M/G Jasper Welch...a brilliant, non-rated, R&D officer. General Welch assigned me the AoA for the GBU-15. Rockwell planned to build the GBU-15 in Duluth, GA for an estimated unit cost of $150K.

I looked for an alternative having the same probability of damage as the GBU-15 but at a lower cost. I decided to test ripple launching (0.4 second between launch) four GBU-12 GP bombs from an F-4. The GBU-12 was a 500-lb, IR guided bomb that cost $20K and was in the inventory. The test was conducted at Nellis AFB, NV by Air Combat Command pilots and consisted of an F-4 designating a target with an IR spot and ripple launching four GBU-12s. After launch, the GBU-12s would fall into trail and follow each other into the target. The targets were an aircraft shelter, a tank, and a simulated hardened command and control center. The test was a success.

I marched into General Welch's office and announced that we should not approve production of the GBU-15 based upon the results of the AoA test. He said "Nicolai, technically you are right...but politically you are wrong." Duluth, GA was an economically depressed area and the GBU-15 production meant jobs. In addition, the chairman of the Senate Armed Services Committee was Sam Nunn...from Georgia.

The IOC for the weapon was 1983 at a unit cost of $245K. Over 2800 GBU-15s have been produced and their performance is spectacular. Note 1) The ripple launch of smaller, cheaper PGMs has become standard practice for the Air Combat Command and has reduced the number of more expensive PGMs employed. Note 2) The AoA means more than just equal performance and the customer can be anybody.

# Chapter 5

## LESSON LEARNED 4 – CHALLENGE THE REQUIREMENTS

"Mother Nature cannot be challenged….But man-made rules can and must be." Clarence "Kelly" Johnson, Lockheed Skunk Works

### 5.1 BACKGROUND

History is filled with cases where the requirements at the start of an aircraft procurement were flawed. In the quotation above the rules that Kelly refers to are the requirements. The requirements should be assumed suspect until an independent analysis proves them to be correct and consistent. The rules could also be ridiculous company procedures that inhibit innovation as well.

Reference [9] reports on a USAF study conducted by a very experienced group of aerospace defense personnel in 2008 on why aircraft programs between 1960 and 2000 experienced cost and schedule growth. The report cites six "seeds of failure" responsible for the cost and schedule growth. These are

1. Inexperienced leadership
2. External interface complexity
3. System complexity
4. Incomplete or unstable requirements at MSB
5. Reliance on immature technology
6. Reliance on large amounts of new software

The report views that the biggest risk of all in undertaking large development programs is to proceed with less than the very best personnel, particularly in the key leadership positions in government and industry.

Seed 1 (inexperienced leadership) and its impact on seed 4 (incomplete or unstable requirements) is the principal reason that the requirements are often flawed. The customer usually publishes a draft RFP, requirements, concept of operations, and system architecture for the sole purpose of getting industry inputs.

The customer requirements, conops, and system architecture must be understood and, if necessary, challenged. The results of the evaluation (along with the assumptions) are shared with the customer prior to RFP release (after RFP release, all communication with the customer must be shared with all the industry competition). This activity is done to make sure that the requirements are understood correctly and are credible. This meeting with the customer must be done with diplomacy and sensitivity. If there are differences, they need to be resolved until a realistic set of requirements can be agreed upon.

If the requirements are not agreed upon and felt to be flawed, then you must walk away from the RFP because you cannot generate a winning design for a set of requirements that you do not believe in. This is a difficult decision for a company to make. In some cases, the company will follow an alternate but very risky path of ignoring the customer's requirements, proposing a design to its requirements and hope that the RFP competition proves it right (this worked for the first example in the next section).

## 5.2 Examples of Flawed Requirements

History is filled with examples of flawed requirements. How industry dealt with the flawed requirements makes interesting stories and generates useful Lessons Learned.

### 5.2.1 TWA Replacement for the Fokker Trimotor F-10A (1932)

Situation: Requirements ignored.

In the late 1920s and early 1930s, the flagship of the TWA commercial transport fleet was the three-engine Fokker F-10A Trimotor (Fig. 5.1). In 1931 an F-10A crashed, taking the life of Knute Rockne, the famed Notre Dame football coach. Inspectors blamed moisture inside the wooden wing that caused the wing structure of the F-10A to separate. The Aeronautics Branch, Department of Commerce (predecessor to the FAA), suspended the airworthiness certificate of the F-10A, grounding a major part of the TWA fleet. In August 1932 TWA issued the specification for a modern luxury transport airplane shown in Fig. 5.2. The TWA requirements were flawed in the sense that they specified the solution as a three-engine aircraft…thus ruling as noncompliant all proposals with other than three engines. Donald Douglas took a risk and offered up a two-engine design…the DC-1. The TWA specs were an extreme challenge for the time, but were met and exceeded by the DC-1, predecessor to the famous DC-3 and World War II C-47. Even though only one DC-1 was built, and 218 DC-2s, the Douglas Aircraft Company turned out 13,300 DC-3s. The fact that DC-3s (Fig. 5.3) are still flying today is a testament to the design genius of Donald Douglas.

Fig. 5.1  Wooden Fokker F10A Trimotor.

### 5.2.2 B-58A HUSTLER (1953)

Situation: Requirements became obsolete during aircraft development... aircraft produced and deployed anyway.

In 1953 the USAF wanted to replace the subsonic, mid-altitude, strategic nuclear bomber B-47 with an aircraft that could overfly the Soviet Integrated Air Defense System (IADS). The IADS consisted of low-frequency (VHF)

---

All metal trimotor monoplane
Payload: 12 passengers
Range: 1000 miles
Crew: 2
Max speed at sea level: 180 mph
Cruising speed at sea level: 145 mph
Landing speed at sea level: 65 mph
Service ceiling: 21,000 feet
Rate of climb at sea level: 1200 fpm
Maximum gross weight; 14,200 lb
Passenger compartment must have ample room for comfortable seats, miscellaneous fixtures, and conveniences.
Airplane must have latest radio equipment, fight instruments, and navigation aids for nigh.

---

Fig. 5.2  TWA specification for a transport aircraft, August 1932.

Fig. 5.3  DC-3…timeless classic.

long-range detection radars, SA-1 surface-to-air missiles, and subsonic fighter interceptors. Intelligence revealed the development of an improved surface-to-air missile, the SA-2 Guideline.

The new bomber was to have a subsonic range of 4000+ nm and a payload of 5 B-61 nuclear weapons (340 kiloton, 700 lb). The government-furnished equipment (GFE) list included a new, sophisticated bomb/navigation system (AN/ASQ-42 and AN/APN-113 radars) with incredible accuracy.

Government operations analysts developed the following requirements for overflying the IADS with the SA-1 SAM: Speed: Mach 2+ Altitude: 60,000+ feet and RF signature: low.

The General Dynamics Corp/Convair Division, Feet Worth, TX designed the delta wing bomber shown in Fig. 5.4, with the following features:

Speed: Mach=2.0
Subsonic ferry range: 4100 nm
Service ceiling: 63,400 feet
Empty weight: 55,000 lb
TOGW: 176,800 lb
Aspect ratio: 2.1
TO W/S: 44 psf
Payload: 5 B-61 nuclear weapons (one in centerline pod, 4 external)

In 1953 the USAF gave Convair a contract to develop the B-58 Hustler. During the development of the B-58, the SA-2 became operational with the Soviet IADS in 1957. U.S. government operations analysis concluded that the CIA U-2A was vulnerable at its maximum altitude and the B-58 was at risk. A U-2A with CIA pilot Gary Frances Powers was shot down on 1 May 1960 over the Soviet Union. (Powers was released two years later.)

## Lesson Learned 4 – Challenge the Requirements

During 1958 the Lockheed Skunk Works conducted an operational analysis to determine the speed, altitude, and RF signature (RCS) needed to defeat the SA-2. Lockheed concluded that Mach 3+ speed and 85,000 foot altitude would provide safe overflight of the SA-2 and Soviet fighters. These parameters became the performance targets for the CIA A-12 Black Bird and the downfall for the B-58 Mach 2+, 60,000+ foot mission profile.

The B-58A first flight was November 1956 and initial operational capability (IOC) March 1960. Initial training flights included both high-altitude overflights and low-altitude terrain following/terrain avoidance (TF/TA) penetration. A new TF/TA, low-altitude bombing system (AN/APN-170 TF/TA radar) was developed.

It was concluded that the B-58A had to penetrate at low altitude in order to survive against the SA-2. The low-altitude penetration reduced the weapon delivery speed to M < 1 and range to less than 1500 nm…essentially a "one-way mission" for the most targets in the USSR. The weapon needed a large drogue chute to give the delivery aircraft separation time to escape the blast effects of the thermonuclear bomb. The aircraft avionics were very sophisticated and were a maintenance nightmare (Lesson Learned 7). The reliability of the low-altitude bomb/nav system was poor, giving it a maintenance index three times that of the B-47. The USAF maintainers called the B-58 a "Hangar Queen," meaning that it spent more time in hangar maintenance than in its mission.

The B-58A was retired in January 1970 after 10 years in service. One hundred and sixteen aircraft were produced.

Fig. 5.4  GD/Convair B-58A Hustler.

### 5.2.3 TFX (1962)

Situation: Requirements challenged by industry but not changed...aircraft built anyway.

In 1962 the USAF wanted a tactical fighter bomber to replace the F-105 Thunderchief and a strategic nuclear bomber to replace the B-58A Hustler. The Navy wanted a deck launch interceptor to replace the F-4 Phantom. Robert McNamara, secretary of defense, decided that the replacement aircraft for the two services would have a common airframe, with the Air Force being the lead acquisition agency. The TFX requirements were (USAF/Navy)

Max speed at sea level: Mach 1.2 (survivability against the SA-3)/Mach 0.9
Max speed at altitude: Mach 2.5/Mach 2.0
Combat Radius (internal fuel): 1160 nm
Payload (internal): two 750-lb M117 bombs and a removable M-61 cannon/
    two AIM-54 Phoenix long-range air-to-air missiles
TO/landing distance: 3000 feet/carrier launch and recovery at 110 knots
Limit load factor: 7.33 g/6.0 g
Cockpit seating: 2 in tandem/2 side by side
Structure: accommodate a Navy 24 f/s sink-rate landing gear

The Mach 1.2 on the deck resulted in a dynamic pressure of 2132 $lb/ft^2$, which was higher than any previous aircraft and forced the inlet structure to be massive due to the high static pressure in the subsonic diffuser. The Navy 110-kt landing speed and Mach 2.0 at altitude forced a variable sweep wing (VSW). The Air Force 3000-ft runway length and Mach 1.2 on the deck and Mach 2.5 at altitude also drove the design to a VSW. The Air Force issued the requirements to industry for comment and was told that

1. The variable sweep wing was high risk. (NASA argued that it was low risk.)
2. The VSW would be 18% heavier than a fixed sweep wing [10], Chapter 20.
3. The inlet would be heavy due to the high dynamic pressure associated with the Mach 1.2 on the deck and the fact that it would have to be a mixed compression inlet for the Mach 2.5.
4. The aforementioned features, plus the 1160-nm combat radius, would result in an aircraft weighing more than 65,000-lb TOGW.
5. The requirements should be relaxed.

The AF team agreed with industry that the requirements should be relaxed, but the DoD refused to change the requirements and the Milestone 1/B RFP was released. All of the proposals were judged unacceptable, but Boeing and General Dynamics were asked to resubmit. The Boeing and GD resubmittal was judged unacceptable and they were asked to resubmit a third time. Finally GD (with

LESSON LEARNED 4 – CHALLENGE THE REQUIREMENTS                                    33

Grumman as an associate for the Navy variant) was selected and the F-111A was built for the Air Force and it weighed 82,800 lb TOGW. The F-111A is shown in Fig. 5.5, with its wings swept forward 16 degrees and aft 72.5 degrees.

The F-111A length and TOGW were unacceptable to the Navy. Grumman could not get a satisfactory design for the Navy variant and the Navy pulled out of the program and contracted with Grumman for the F-14 Tomcat.

General Dynamics had significant development problems due to the flawed assumption that the critical technologies were low risk. The inlet/engine (TF 30 turbofan) configuration had compatibility problems due to the high inlet flow distortion levels. The VSW structural fatigue analysis was immature, leading to wing pivot/wing carry through structural failure.

Fig. 5.5   F-111 with wings swept forward (below) and swept aft (above).

General Dynamics produced 463 F-111A/C/D/E/F aircraft for the USAF and foreign countries and 76 FB-111A to replace the B-58A. The aircraft featured a variable sweep wing (heavy), a variable geometry, mixed compression inlet (heavy), and a sophisticated low-altitude/terrain following subsystem (expensive and poor reliability). It is pretty much agreed that the F-111 was designed to flawed requirements. David S. Lewis (Chairman and CEO of General Dynamics) had the following to say about the F-111 in his 67th Wilbur and Orville Wright Memorial Lecture to the Royal Aeronautical Society, 7 December 1978: "The F-111 is truly a remarkable aircraft but unfortunately is very heavy, expensive and has poor reliability. Had more thorough tradeoffs been made at the outset, it is almost certain that a decision would have been made that a sea-level dash speed of Mach 0.8 or 0.9 would have had an acceptably high probability of survival. The aircraft would have been smaller, simpler and much cheaper. The USAF could have afforded more and the effectiveness of the overall inventory would have been much higher for the dollars expended."

Reference [11] summarizes the mistakes made in the requirements definition, requirements management, and design tradeoff studies. "The ill-conceived, difficult-to-achieve requirements and attendant specifications made the F-111 system development extremely costly, risky and difficult to manage. The government F-111 System Engineering managers were not allowed to make the important tradeoffs that needed to be made in order to achieve an F-111 design that was balanced for performance, cost and mission effectiveness (including survivability) and the attendant risk and schedule impacts."

The problems with the TFX can be directly attributed to the experience of the DoD personnel generating the requirements and to the restrictions imposed by the SecDef for a common airframe development program.

The author was an AF captain assigned to the Developmental Plans Directorate, Aeronautical Systems Division (ASD/XP), W-PAFB, Dayton, OH in September 1961. He was part of an AF team charged with developing the requirements for the TFX weapon system. NASA took a keen interest in the requirements and pushed for the Mach 1.2 on the deck as justification for their research investment in variable sweep wings. NASA convinced the AF and Navy management that VSW was mature and low risk.

### 5.2.4 RULE 2 – ALL FIGHTERS WILL HAVE AN INTERNAL GUN

Situation: The F-4 was initially designed without an internal gun and many MIG kill opportunities were missed.

The F-4 Phantom was designed in the 1950s for the U.S. Navy as a deck launch interceptor by McDonnell Aircraft. It was deployed in 1960 by the Navy with the RF AIM-7 BVR (beyond visual range, operational range 20+ miles) missile and the IR AIM-9 (0.6 to 22 mile operational range) missile. Because of the BVR interceptor mission, the fighter did not have a gun for close-in combat. The USAF deployed the Navy F-4C in the SEA Vietnam War for air superiority (protecting the F-105s) and ground attack. In 1967 the F-4Cs were fitted with an RF AIM-4 Falcon missile. The AIM-4 had poor performance and reliability and was removed the following year, keeping the AIM-7 and AIM-9. The USAF pilots were complaining about the missed opportunities for MIG kills by not having a gun [12]. In 1967 the F-4C/Ds were fitted with a 20-mm external gun pod on the inboard wing pylon. The gun pods gave the pilots the opportunity for close-in gun kills, but the pylon vibration greatly reduced the gun pod's accuracy. In 1973 (the last year of the Vietnam War) the F-4Es were delivered with an internal 20-mm Vulcan cannon that was very effective. In the SEA East Asia conflict, the F-4s killed 107 MIGs while losing only 33 in air combat. The summary of MIG kills is as follows:

| MIGs killed by… | kills |
|---|---|
| AIM-7 missile | 50 |
| AIM-9 missile | 31 |
| AIM-4 missile | 4 |
| 20-mm gun pod | 10 |
| M-61 20-mm cannon | 6 |
| Maneuvering tactics | 5 |

If the internal gun could have been available at the beginning of the SEA conflict, it is estimated that the MIG losses to guns could have exceeded the kills due to the missiles.…Waiting to get an RF missile firing solution or an IR lock-on resulted in many missed opportunities [12].

### 5.2.5 ADVANCED TACTICAL FIGHTER (1985)

Situation: Requirements challenged by industry and changed.

We discussed the 1986 Advanced Tactical Fighter (ATF) in the previous chapter as an example of listening to the customer. But the story didn't start in 1986…it started a year earlier in 1985 with the publication of the RFP for the ATF. In 1985 the USAF issued a Milestone B RFP for a new air-to-air advanced tactical fighter to replace the F-15. The RFP requirements called for "super maneuver" and "super cruise" (the ability of the aircraft to cruise at supersonic speeds without afterburners)…and a modest signature level. Industry challenged the requirements and argued that the radar cross section (RCS) should be lowered to enable the ATF to be a true twenty-first-century fighter. The USAF agreed and in November 1985 they modified the Milestone B RFP by lowering the RCS requirement to a super stealth level and extended the proposal due date to February 1986. The Lockheed entry was the very low observable configuration shown in Fig. 4.4. The Lockheed Skunk Works and Northrop Aircraft Division won contracts to build two prototypes each (the YF-22 and YF-23 respectively) and have a fly-off.

### 5.2.6 JASSM (1995)

Situation: Requirements challenged by the Lockheed Martin Skunk Works and changed.

In the spring of 1995 the Department of Defense (DoD) canceled the stealthy air-launched cruise missile named Tri-Service Standoff Attack Missile (TSSAM/AGM-137, discussed in Section 8.2.1) because of excessive unit cost. The mission need for the TSSAM still existed and so a draft RFP was issued to industry in the fall of 1995 for Joint Air-to-Surface Standoff Missile (JASSM). The unit cost requirement was $400,000, the same as that for TSSAM, but the all-aspect stealth requirement was reduced to front sector only. The USAF concluded that the only way to meet the cost target was to ask for a derivative of an existing cruise missile (forcing the change in the signature requirement) and so the Milestone B RFP specified a derivative missile. Lockheed Martin questioned the requirements. The Skunk Works convinced the USAF that they could have the same performance and unit cost as that of TSSAM with a clean sheet design. The JASSM RFP was reissued in the winter of 1996 asking for a derivative or a clean sheet design with front-sector-only stealth but with the same cost target. Lockheed Martin and McDonnell Douglas both won contracts with

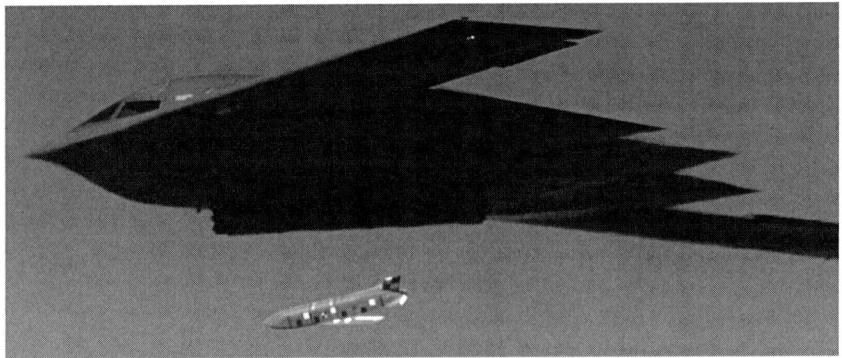

Fig. 5.6   JASSM launch.

clean sheet designs for the follow-on risk-reduction phase. Lockheed Martin won the downselect for the production phase of over 2400 AGM-158 cruise missiles (shown in Fig. 5.6 being launched from a B-2A). Lockheed Martin's strategy for winning the production contract is discussed in Section 7.2.2.

> Once again the incumbent did not win. The Skunk Works had no existing missile to modify so it was forced to rethink how best to compete against existing missiles. Intensive trade studies convinced the customer that a clean sheet design could meet the target cost and so the RFP was modified, and the rest is history.

# Chapter 6

# LESSON LEARNED 5 – POPULATE THE DESIGN TEAM WITH LEFT AND RIGHT BRAIN ENGINEERS

## 6.1 BACKGROUND

The brain has two lobes…the left brain for analytical, critical, and judgmental thinking and the right brain for creative and innovative thinking. Engineers are predominately left brain and so populating a team with left brain engineers should be easy. Finding right brain engineers will be harder as right brain people usually do not go into engineering. With practice, engineers can learn to flip their brains from left to right and back, but they will never be as creative as truly right brain individuals. The conceptual design phase requires both types of thinking but fortunately not at the same time.

Referring to Fig. 1.1, we observe that different design activities require left brain thinking and others require right brain. Setting up the ground rules, evaluating the candidate concepts, assessing risk, and selecting a baseline are all left brain activities because the laws of physics have to ultimately prevail. Developing the candidate concepts is clearly a right brain activity because we are looking for creative and innovative ideas. This activity must be a nonjudgmental brainstorm activity to ensure that a gem (such as a DC-3—it had only two engines) does not get overlooked. "Out of the box" thinking is encouraged and "rush to judgment" on the unconventional concepts is discouraged. Having right brain inputs at this point in the conceptual design phase is critical since success or failure of this phase depends upon having a stable of candidate concepts to select from.

## 6.2 EXAMPLES FROM THE PAST

### 6.2.1 THE JOINT STRIKE FIGHTER (JSF) COMPETITION (1996)

Boeing's concept for the JSF competition was the vectored thrust vertical lift system. This concept was technically sound and low risk, having been the vertical lift system on the AV-8 Harrier. The JSF engine was the PW F119 after-burning turbofan with ~ 32,000 lb of vectored thrust available for VTOL at sea level on a 75-deg day. During hover, the hot gas from the vectored thrust migrated forward on both the AV-8 and X-32B and was ingested into the engine inlet, causing a sag in the available thrust. The vectored thrust concept shown on the left in Fig. 6.1 on the Boeing X-32 prototype was a left brain

**Fig. 6.1** JSF prototypes showing left brain X-32B and right brain X-35B thrust concepts.

approach and was marginal in meeting the very challenging STOVL requirement for the USMC variant of the JSF.

However, the Lockheed Martin Skunk Works went with the unproven, high-risk, right brain concept of a shaft-driven lift fan (SDLF) for the front post and vectored thrust for the rear post and roll jets. Most of the risk was in developing a gear box to drive the SDLF, a clutch, and a functional light-weight fan. At sea level, the X-35B was able to generate 39,100 lb of vertical lift distributed as follows: 16,411 lb from the aft three bearing swivel nozzle, 3607 lb from the wing-tip roll jets, and 19,082 from the lift fan. The lift fan achieved an augmentation ratio of about 1.6. The downwash from the lift fan (located just behind the cockpit and inlets) was a cool jet of air which formed a "fence," which prevented the hot gas from the nozzle and roll jets from migrating forward and being ingested into the engine. The SDLF plus vectored thrust of the X-35B (shown in Fig. 6.1) had considerable margin (over 22%) during vertical landing over the vectored thrust concept of the X-32B and won the JSF competition. The aircraft is now in production and will be in service for many decades in the Air Force, Navy, and Marine Corps as the F-35A, F-35C, and F-35B respectively.

### 6.2.2 DARPA NATIONAL AERO-SPACE PLANE (NASP)/X-30 PROGRAM

In 1983 Tony DuPont offered an unsolicited proposal to DARPA to design a single-stage-to-orbit (SSTO) vehicle. It was to be a hypersonic vehicle powered by a hybrid integrated engine of scramjets and rockets. The idea was funded from 1983 to 1985 and was called Copper Canyon. DuPont's concept was clearly a right brain idea and was built upon overly optimistic assumptions and performance goals.

But the excitement of SSTO captured the nation and in 1985 President Reagan introduced the National Aero-Space Plane (NASP) program. The program was to conclude with the fabrication of a demonstrator vehicle termed the X-30. The program was canceled in 1993 due to both weight and budget growth coupled with, what the aerospace community viewed to be, a technological overstretch. Quite simply the scramjet performance and weight goals were not going to be met.

# Chapter 7

## LESSON LEARNED 6 – THE INTEGRATED PRODUCT TEAM (IPT) WORKS – USE IT

### 7.1 BACKGROUND

The Integrated Product Team is a group of project or program people, each representing a discipline (such as aerodynamics or flight controls), a function (electrical or hydraulic power), a subsystem (propulsion or fuel system), or a system (airframe or mission systems). The membership on the IPT constitutes every person in that discipline, function, subsystem, or system that will ever touch the product…all the stakeholders. The IPT reviews and approves every change to the product with every member having an equal voice. For example, the design IPT cannot close on the design until the manufacturing member approves that it can be built and the maintenance representative agrees that it can be maintained. The IPT became an accepted element of best practice in the 1970s and a required activity in the Federal Acquisition Regulations (DoD 5000.1) in the 1990s. Before the 1970s, design drawings were often thrown over the transom to the surprise of the shop, as depicted in Fig. 7.1, resulting in parts that were difficult or expensive to fabricate.

All members of the IPT have a sense of product ownership that is required for a quality, on-time delivery of a product at the lowest cost. This concept greatly increases the depth and breadth of the trade studies. All proceedings and decisions are documented and made part of the products file.

### 7.2 EXAMPLES FROM THE PAST

#### 7.2.1 F-4 EJECTION SEAT

The maintenance records on the early F-4s reported that the rear ejection seat was a maintenance nightmare [13]. For about every three sorties the rear ejection seat had to be removed and then replaced, costing 2.3 maintenance man-hours. Upon close examination of the maintenance event, it turned out that the ejection seat was not the problem. The McDonnell designers had located some communication avionics in an empty cavity beneath the rear seat. Every time the communications gear needed adjusting (which was about every three sorties), the rear ejection seat had to be removed and then replaced. This poor design packaging added 2.3 maintenance man-hours to the aircraft turnaround time just to gain access to the avionics equipment.

**Fig. 7.1** ...throwing the design over the transom.

If there had been an IPT of record in place during the design of the F-4, chances are that this design glitch would not have occurred. The operation and maintenance (O&M) representative on the IPT would not have approved the design.

### 7.2.2 O&M OF LOW-OBSERVABLE AIRCRAFT

The introduction of the stealth fighter (F-117A) and stealth bomber (B-2A) into the USAF air combat fleet required some unique design decisions made early in the design cycle. The low RCS features of the aircraft required special materials and coatings that were fragile and a nightmare to maintain. A wrench dropped on the wing leading edge could poke a hole in the fragile skin material that required special equipment and hours to repair. The Special Technology (LO) engineers and the maintenance personnel were members of the airframe IPT from the very beginning of the design to ensure that the materials and coatings could withstand the operational environment and could be maintained in the field. Some materials were not suitable and were voted off the design. Field repair techniques had to be developed and special equipment made a part of the ground support equipment (GSE). The O&M data below shows a steep learning curve in maintenance man-hours per flight hour from initial deployment (IOC) to mature operation (approximately 35% of the maintenance man-hours are due to the LO materials).

| Aircraft | Date | MMH/FH |
|---|---|---|
| F-117A | 1983 (IOC) | 113 |
| F-117A | 2003 | 45 |
| B-2A | 1997 (IOC) | 124 |
| B-2A | 2002 | 51 |

## Chapter 8

# LESSON LEARNED 7 – KISS (KEEP IT SIMPLE STUPID)…AS LONG AS YOU CAN

## 8.1 BACKGROUND

This should be a "no brainer" but it's not. Engineers must be reminded to keep their designs simple. The young left brain engineers strive for the challenging and complicated design. They love to show off with a sophisticated solution and then write a paper and present it at a technical conference to impress their peers. The older engineers do not have this problem as much because they have been beaten over the head by the shop people trying to build their design and by the crew chief trying to maintain it. This difference in design philosophy is shown facetiously in Fig. 8.1 as two solutions for the design of a rope swing. The complex design will always cost more and have less reliability.

Sometimes the simple solution does not meet the requirement so a complex design or new technology is necessary to meet the performance target, BUT it must "buy" its way onto the airplane and be approved by the airframe IPT with all members buying in. The trade-off analysis must show that the aircraft is more cost-effective with the more complex design or the new technology. A perfect example is the X-35B with its more complicated SDLF (see Section 6.2.1). The all-vectored thrust concept on the X-32B was clearly a simpler approach, but it did not have the performance margin to meet the very challenging JSF STOVL requirements.

Sometimes the problem needs technology that is not mature or available. This was the case for the DARPA NASP/X-30 airplane discussed in Section 6.2.2.

## 8.2 EXAMPLES FROM THE PAST

### 8.2.1 TSSAM/AGM-137 AIR LAUNCH CRUISE MISSILE

In 1986 the DoD began to develop a stealthy cruise missile for the U.S. Air Force, U.S. Navy, and the U.S. Army called Tri-Service Standoff Attack Missile (TSSAM). The Navy was the acquisition lead agency. The Air Force and Navy would use the air-launch variant AGM-137A launched from the B-52H, F-16C/D, B-1B, B-2A, A-6E, and F-18C/D. The Army would use the

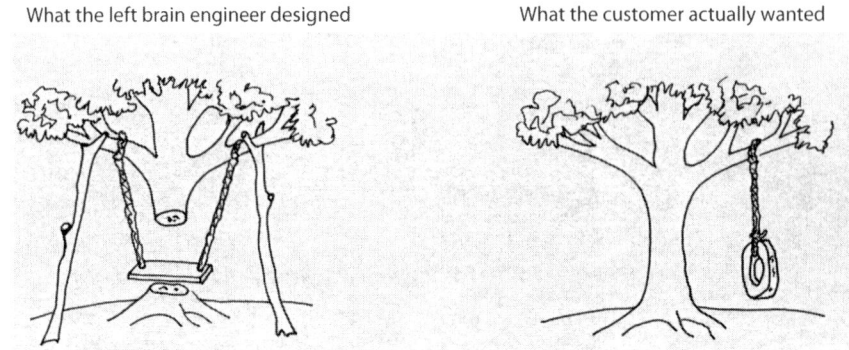

Fig. 8.1  Remember…KISS.

ground-launched variant MGM-137B launched from the Multiple Launch Rocket System (MLRS). The AGM-137A would carry a 1000-lb class unitary conventional warhead, and the MGM-137B would carry a 1000 lb of Combined Effects Bomblets (CEB). The Army pulled out of the program in 1993 and developed their own MGM-140 ATACMS (which later became the MGM-164 ATACMS II).

The missile configuration, length (14 ft) and weight (2000 lb), was driven by aircraft rotary launcher and wing pylon limits (see Fig. 8.2). The missile was powered by a single Williams F122 turbofan and featured all-aspect stealth, and an autonomous guidance and target recognition system. The AUP (all–up unit price) target was $400,000 in 1986 dollars. Northrop Aircraft Division, Hawthorne, CA was the contractor.

The Northrop engineers forgot they were building a bomb and designed a very sophisticated and elegant airframe and subsystems, leading to technical problems during development and flight testing. Each launch aircraft required a unique set

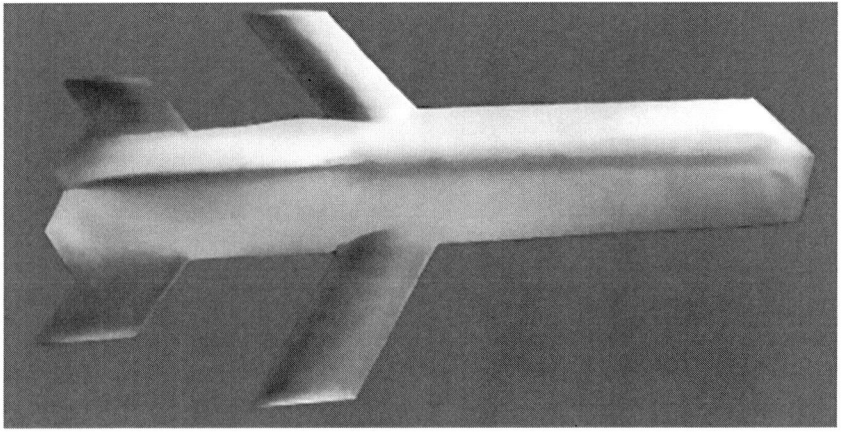

Fig. 8.2  Northrop TSSAM/AGM-137A.

of strakes affixed to the missile aft body for safe separation during the unpowered launch and all eight had to be packaged in the AGM-137 all-up round container. The program was canceled in 1994 when the AUP exceeded $2M.

### 8.2.2 JASSM/AGM-158 AIR LAUNCH CRUISE MISSILE

The requirement for the TSSAM was still valid and so the DoD started the Joint Air-to-Surface Standoff Missile (JASSM) in 1995 with the Air Force as the lead acquisition agency. The JASSM requirements were the same as for TSSAM except the stealth was relaxed to front sector only. The target AUP remained at $400,000 in 1986 dollars and was the design driver. The payload was the WDU-42B (J 1000) 1000-lb penetrating warhead. The engine was changed to the lower-cost Teledyne CAE J402 turbojet. The F-15E was added to the list of launch platforms and the A-6E dropped. The JASSM was initially to be a derivative of an existing cruise missile but was changed to a clean sheet design in 1996 (see Section 5.2.6). Lockheed Martin/Missiles and Fire Control, Orlando, FL and McDonnell, St. Louis, MO were awarded 24-month program definition and risk-reduction contracts in June 1996.

The Lockheed Martin engineers studied the TSSAM overrun and devised a strategy to control costs. The strategy was simplicity in design, minimizing the parts count, single points of failure, and a structure designed for multi-function. The horizontal tail was eliminated with pitch control coming from the wing flaps. This one design feature saved two actuators and two surfaces for a cost saving of ~ $50K in 1986 dollars (see Fig. 8.3). The airframe was made from room cure advanced composites…the fuselage was fiber wound and the wing and vertical surface used surfboard fabrication techniques.

In April 1998 Lockheed Martin was awarded the engineering and manufacturing development (EMD) contract. The AGM-158A was certified for operational use in October 2003. The current USAF requirement is for up to 4900 cruise missiles.

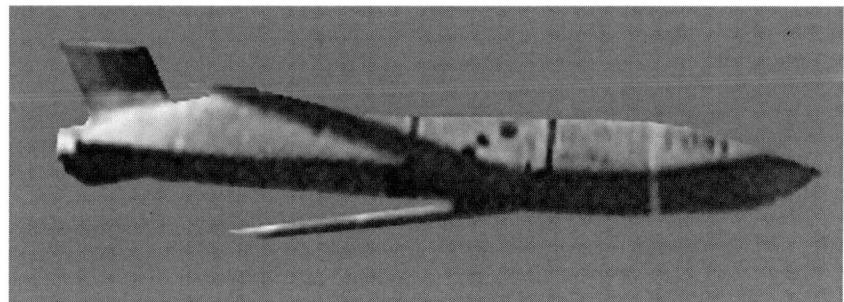

Fig. 8.3   The Lockheed Martin JASSM/AGM-158 cruise missile.

# Chapter 9

## LESSON LEARNED 8 – WILLOUGHBY TEMPLATES WORK…USE THEM

### 9.1 BACKGROUND

In 1982 the DoD commissioned the Defense Science Board (DSB) to address the issues of transitioning from development to production. Willis J. Willoughby, Jr. was the chairman of the DSB task force. This task force came out with compilation of best practices to shorten the time and reduce the cost of developing and deploying weapon systems. The results were published as DoD Directive 4245.7 "Transition From Development to Production," September 1985.

In March 1986 the Navy transformed the results of DoD Directive 4245.7 into a series of templates and published them in NAVSO P-6071. These templates take each function of the transition (design, test, production, facilities, logistics, and management) and breaks them into subfunctions. For example, design is broken into design requirements, trade studies, design policy, design process, design analysis, etc. The subfunctions are discussed in detail and conclude with a checklist of questions to aid the program manager (government and industry) in deciding if the subfunction has been completed. Taken together, the templates give a good indication of the readiness of the product to transition from development to production.

The transition functions covered by the templates in NAVSO P-6071 are shown in Fig. 9.1. Each function is discussed and a comparison made with the current approach and best practices. Things to look out for (TRAPS) are identified for each function and a checklist is developed. As an example, Fig. 9.2 shows the TRAPS identified for the design requirement function and Fig. 9.3 the checklist recommended for moving forward.

The Willoughby templates are useful in the early conceptual design phase to impress upon the project personnel that it is never too early to plan for the production of the design. The templates hammer home the following:

1. The design must be taken serious with the full intention that it will be built.
2. Quality cannot be manufactured in…it must be designed into the product.

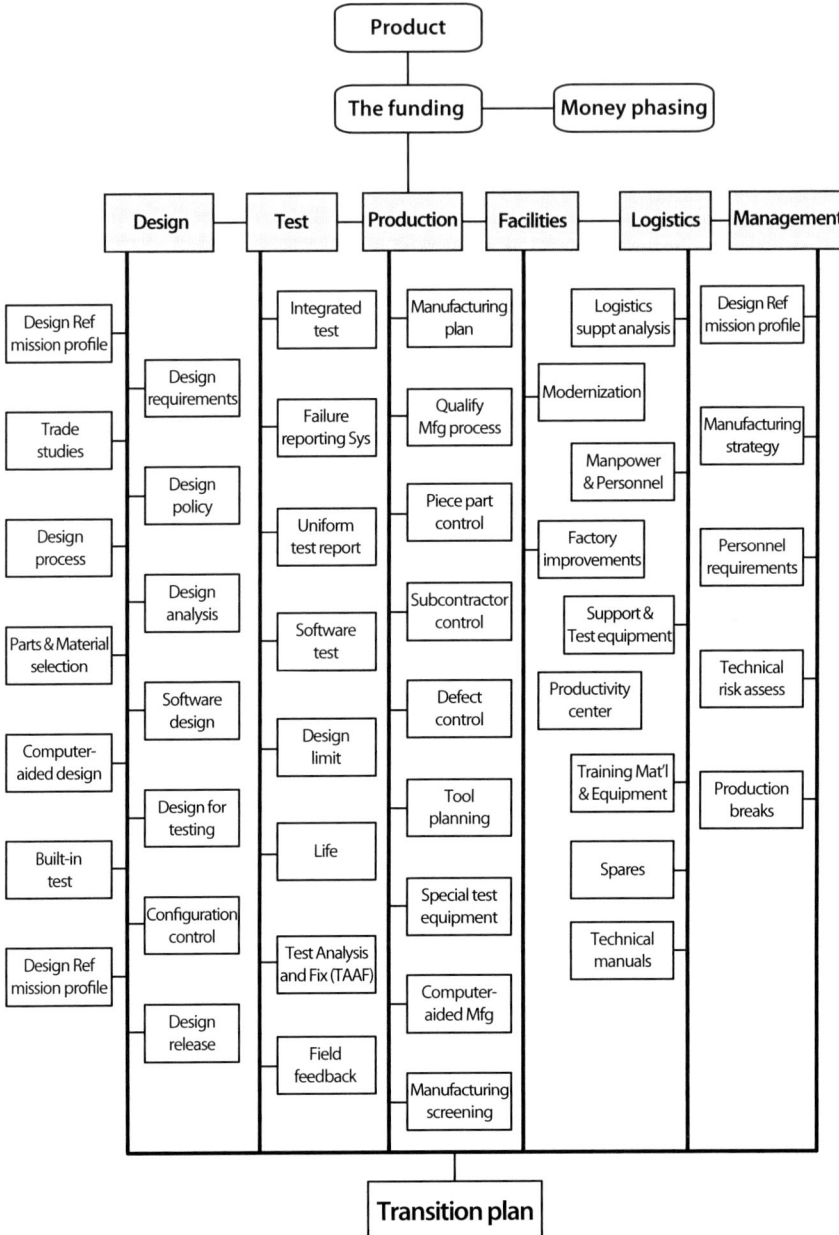

**Fig. 9.1  The transition functions covered by the Willoughby templates.**

## TRAPS

1. Operational requirements are stated as design requirements.
2. Management policy stresses on program and delivery schedules.
3. Latest technical developments are used as a basis for design requirements.
4. Detailed design requirements evolve with design effort.

**Fig. 9.2** The TRAPS identified for the design requirement transition function.

3. Reliability, maintainability, and supportability (RM&S) cannot be manufactured in...it must be designed into the product.
4. Producibility cannot be worked out on the shop floor...it must be designed into the product.

### 9.2 EXAMPLES FROM THE PAST

#### 9.2.1 F/A-22 TRANSITION

In 2003 Lockheed Martin Aeronautics was having trouble transitioning the F/A-22 from development to production. In February of that year, Air Force

## Design Requirements

### Checklist

✓ Have mission needs been interpreted and specified as measurable design parameters?

✓ Have system design requirements been specified for, allocated to, and understood by each responsible design engineer?

✓ Have relevant design requirements been flowed down to subcontractors?

✓ Have detailed design requirements been specified in the RFP?

✓ Is Inherent Availability (Ai) used as a design requirement?

✓ Have design requirements been frozen at Milestone II?

**Fig. 9.3** The Design Requirement Checklist for moving forward.

Secretary James Roche admonished Lockheed Martin [13]. He said "Lockheed Martin did not pay the attention it should have to classic transition issues... Willoughby templates guys, Willoughby templates, hello, hello." Copies of the templates were issued to all F/A-22 program personnel and made required reading. The F/A-22 Raptor was certified operationally ready in December 2005. In 2006 the F/A-22 won the Collier Trophy.

# Chapter 10

## LESSON LEARNED 9 – PLAN FOR VARIANTS

### 10.1 BACKGROUND

There are three things in life that are for certain…death, taxes, and requirements change. The first two are a nuisance, but the third can really spoil your day. The requirements drive the design of your aircraft and when they change you are left with a relic of yesterday's requirements. The result is usually a reduction in the planned production number of the baseline aircraft. As soon as your design is complete on your contracted effort (usually called "design freeze"), you should start designing variants of the baseline design that could accommodate changes to the requirements. Because the requirement changes are generally undefined at this point in time, you should design in a degree of robustness.

**Rule 3 – The requirements will change…deal with it**

The baseline design must be able to "morph" into a variant that can do a different mission as the requirements change. When the variants start to emerge, the design will be given the coveted label of "multi-role," which is a good thing, as it denotes a design that is cost-effective. Even more important is that it denotes a large production run.

Immediately the question arises, "How do you optimize the aircraft design for the basic set of requirements and expect it to do well meeting different and undefined requirements?" The answer is you don't build the optimum design…but rather you meet the baseline requirements and design in a robustness to accommodate the variant features to do other missions. This takes considerable skill and insight on the part of the chief designer to decide how much robustness can be incorporated in the baseline design and still win the competition. Adding robustness always increases the weight and cost of the baseline design.

As was mentioned earlier, our profession is one of building airplanes for fun and profit. Chapter 3 discussed the aspect of having fun and this chapter will discuss the aspect of making a profit. The profit comes about by producing and selling airplanes for more than it costs to make them. The more airplanes you make and sell, the bigger the profit. The secret to selling lots of airplanes is to have many variants of the basic airplane. Having variants will keep the line of business (LOB) sold. Each variant would be a modification of the basic airframe but tailored to fit a different mission. The variants would be the basic aircraft with a simple modification to the airframe such as a plug in the fuselage to accommodate a second crew member, or more fuel, or a different engine or payload.

The project design team should always plan for design variants in the future and do the design work to quantify the performance and cost for each variant. Always have an engineering change proposal (ECP) ready to show to the military customer or a marketing brochure for the commercial customers. It is always an advantage to be able to show that your design has growth potential and a long service life in meeting a lifetime of changing requirements.

## 10.2 Multiple Variant Designs

### 10.2.1 F-4 Phantom II – 5195 produced worldwide

F-4A/B Navy Deck Launch Interceptor (AIM-7 Sparrow RF and AIM-9 Sidewinder IR missile, no gun)
F-4C/D Air Force Air Superiority (AIM -7 and -9 missiles, bombs, no gun)
F- 4E/F Air Force and USMC Strike (AIM -7 and -9 missiles, bombs, 20-mm external gun pods or 20-mm M-61 internal cannon)
F-4G Electronic Counter Measure/Jammer RF-4 Reconnaissance

### 10.2.2 F-15 Eagle – 1198 produced for the USAF

F-15A/B/C Single-Seat Air Superiority
F-15D/E Two-Seat Strike

### 10.2.3 F-16 Fighting Falcon – 4540 produced worldwide

F-16A Air Force Single-Seat, Day Air Superiority
F-16B Air Force Single-Seat All-Weather Air Superiority
F-16C/D Air Force Multi-Role Fighter (A-A and Air-to-Ground)
F-16N U.S. Navy Adversary and Threat Simulation

### 10.2.4 F-18 Hornet, Super Hornet, and Growler – 2109 produced for the U.S. Navy/USMC

Hornet – Derived from YF-17, F 404 engine (1480 produced)
F/A-18A Single-Seat Fighter/Attack
F/A-18B Two-Seat Fighter/Attack
F/A-18C Single-Seat Air-to-Surface Strike
F/A-18D Two-Seat Strike and Recce
Super Hornet – Derived from the Hornet, fuselage stretched 34 inches aft of the crew station, wing area increased by 25%, some low-observable features, F 414 engine (35% more thrust), extensive avionics upgrade (515 produced)

F/A-18E Single-Seat Multi-Role
F/A-18F Two-Seat Multi-Role
Growler – Derived from Super Hornet, 114 produced
F/A-18G Electronic Counter Measure/Jammer

### 10.2.5 BOEING 777 (TRIPLE 7) – 1372 PRODUCED THROUGH FEBRUARY 2016

Variants are -200, -200ER, -200LR, -300, -300ER, freighter, -800 and -900 accommodating different numbers of passengers and seating arrangements, and range.

### 10.2.6 AGM-86 ALCM (AIR LAUNCH CRUISE MISSILE) – 2004 PRODUCED FOR THE USAF

AGM-86A Baseline design, nuclear warhead AGM-86B Nuclear warhead (W-80) and TERCOM (terrain contour matching) guidance AGM-86C Conventional blast/fragmentation warhead (1200 lb) with GPS guidance AGM-86D Conventional penetrator warhead (1200 lb) with GPS guidance

## 10.3 SINGLE-VARIANT DESIGNS

Some missions are so specialized that the baseline design has to be optimized to the extent that the design cannot afford any robustness.

### 10.3.1 A-10 THUNDERBOLT II – 716 PRODUCED FOR THE USAF

A-10A is a single-seat, subsonic close air support, (CAS), attack fighter. Aircraft was tightly optimized around the GAU-8 30-mm rotary cannon and the low speed/altitude CAS mission.

### 10.3.2 AGM-129 ADVANCED CRUISE MISSILE – 461 PRODUCED

AGM-129A ACM was a stealthy, subsonic, terrain following, air-launch cruise missile with a nuclear warhead (W-80). The aircraft was planned to replace the AGM-86, but the end of the Cold War changed the plan (see Section 10.4).

### 10.3.3 B-58 HUSTLER – 116 PRODUCED FOR THE USAF

B-58A was a high-altitude, supersonic, strategic nuclear bomber, but the introduction of the Soviet SA-2 surface-to-air missile forced the aircraft to low-altitude penetration (discussed in Section 5.2.2).

### 10.3.4 F-117 NIGHTHAWK – 64 PRODUCED FOR THE USAF

F-117A was a single-seat, subsonic, low-observable, ground attack fighter. The aircraft was very specialized and production was limited to 5 test and 59 operational. The aircraft was operational in 1983. It was retired in 2008 because of the IOC of the F/A-22 and development of the F-35.

### 10.3.5 B-2 SPIRIT – 21 PRODUCED FOR THE USAF

B-2A is a very specialized subsonic, low-observable, conventional/ nuclear bomber. Production was limited to 21 aircraft because of their cost ($929M in 1997 for each aircraft plus spare parts, GSE, and software support).

### 10.3.6 F/A-22A RAPTOR – 187 PRODUCED FOR THE USAF

The F/A-22A is a single-seat, twin-engine, all-weather, stealth-tactical fighter. It was designed for air superiority. It is currently unmatched in the air-to-air role. It is capable of ground attack, electronic warfare, and ISR data fusion.

The original production run was to be 753 aircraft, but the number was reduced in increments down to the final number of 187 by President Bush in 2004. The reasons given for the reduced production run were the high cost of the aircraft ($361M each in 2006), a lack of a clear air-to-air mission due to the delays in the Russian and Chinese fighter programs, a ban on exports, and the development of the more versatile and lower-cost F-35.

## 10.4 AGM-86 ALCM AND AGM-129 ACM

In 1974 the USAF contracted with Boeing to develop a long-range air-launch cruise missile with a nuclear warhead (W-80) as a U.S./NATO strategic deterrence against the Soviet Union during the Cold War. The AGM-86 ALCM (shown in Fig. 10.1) was produced from 1980 to 2001 in three variants: the AGM-86B with the W-80 nuclear warhead (1715 produced), the AGM-86C with a 1200-lb conventional blast/fragmentation warhead (239 produced), and the AGM-86D with a 1200-lb conventional penetrating warhead (50 produced). The AGM-86 weighs 3200 lb and is launched from a B-52. For survival against the enemy ground and air defenses, the AGM-86 flies a low-altitude TF/TA (terrain following/terrain avoidance) mission profile at Mach 0.7 and 250-ft AGL (above ground level). The range for the AGM-86B is 1500+ nm, and for the AGM-86C/D it is 680 nm. The AGM-86B was operational in 1982 and the AGM-86C/D in 1991.

Fig. 10.1  AGM-86 ALCM.

The decision for forward sweep instead of aft sweep was due to several factors. The wing leading edge RCS spike bounced off the forward fuselage (absorbing energy due to the local RAM coating) and reflected away from the threat radar. The wing carry through used the same structure as the forward carriage lug (saved weight and cost). The X-29 forward sweep wing demonstrator was being developed in DARPA at the same time and provided advanced composite tailoring information to minimize the wing weight increase due to aeroelastic divergence

In the mid-1970s it became clear to the USAF that the AGM-86 was going to have difficulty surviving against the Soviet SAMs and airborne interceptors. The Defense Advanced Research Projects Agency (DARPA, the government Skunk Works) began design studies of a very low-signature cruise missile under the code name TEAL DAWN Advanced Cruise Missile (ACM) to survive against Soviet air defenses.

This author was a colonel in the USAF assigned to DARPA and was the TEAL DAWN program manager. The mission requirements for launch weight, range, payload, and mission profile were the same as for the AGM-86. Conceptual and preliminary design contracts were awarded to multiple aircraft companies with General Dynamics/Convair Division selected in 1981 as the winner of the TEAL DAWN program. In 1982 the USAF awarded Convair a contract to develop their TEAL DAWN design as the AGM-129 ACM. The Convair design (shown in Fig. 10.2) featured a chiseled nose for +/- 25-deg forward sector low RCS, a flush inlet to eliminate the inlet RF reflection back

Fig. 10.2 AGM-129 ACM.

to the threat, forward swept wings to reflect the wing leading edge RCS spike away from the threat, and an upper surface ramp over the nozzle to hide the exhaust IR from look-down airborne interceptors. The plan was to have the AGM-129 replace the 1715 AGM-86Bs.

The Berlin Wall was a symbol for the Cold War as it divided Germany into an East and a West. The wall was erected in 1962 and it separated families and interrupted commerce in the Father Land. In 1989 the German people were fed up and pressed for reunification of Germany and brought the Berlin Wall down in November 1989. Two years later in January 1992 President Bush declared the Cold War over and cut the procurement of the AGM-129A to 640…shortly thereafter the production was reduced to 460 where it remained until retirement in 2012.

The AGM-129 first flew in 1985. The Convair engineers had two assignments: 1) transition the development to production (Lesson Learned 8), and 2) develop conventional warhead variants. The engineers had trouble with assignment 1 because of hardware quality problems, flight test mishaps, and a machinists strike in 1987. When production stopped in 1991 they did not have a credible design for a conventional warhead variant. The USAF tried to get funding to develop a conventional variant, but the GAO did not see that as a good way to spend sparse military dollars…and besides, the USAF had the AGM-86C/Ds. The high-technology, low-signature AGM-129A was operational from 1991 until its retirement in 2012, serving without a mission as a Cold War relic…an ACM was never fired in anger.

General Dynamics/Convair sold the AGM-129 LOB to Hughes in 1992 who sold it to Raytheon Missiles in 1997.

**So, the message is…**

- Requirements will change.
- Having variants will help keep the LOB sold.
- Out-hustle the competition…pursue variants with a vengeance.

# Chapter 11

## LESSON LEARNED 10 – EXHIBIT NYMPHOLEPSY

### 11.1 BACKGROUND

Nympholepsy –The yearning for the unachievable.

In 1959 the conventional wisdom was that a Mach 3+ airplane was too hard. In 1961 the conventional wisdom was that putting a man on the moon by the end of the century was too hard. In 1975 the conventional wisdom was that a tactical fighter with an RCS the size of a ball bearing was too hard. Fortunately we had leaders who would not settle for the conventional wisdom.

"It is hard to soar like an eagle when you work with turkeys"…or work with people of conventional wisdom (origin of quotation is unknown).

### 11.2 EXAMPLES OF NYMPHOLEPSY

1. On 29 August 1959 President Eisenhower approved a CIA contract with the Lockheed Skunk Works to develop a Mach 3 high-altitude reconnaissance aircraft to replace the U-2A. The program was called the A-12 Archangel and had a crew of one. The program was highly classified with a code name OXCART. The development included pioneering titanium structure, unique dual cycle (turbine and ramjet) engines, special high-temperature fuel and lubricants, stealth technology, and much more. The first flight for the CIA was 22 April 1962. Fifteen A-12s were produced for the CIA. The USAF contracted with the Skunk Works in February 1963 for a two-seat variant of the A-12 (SR-71). Thirty-two aircraft were produced for the USAF to do ISR. The aircraft first flew in December 1964, became operational in 1966, and was retired in 1998. The operational speed for the SR-71 Blackbird (shown in Fig. 11.1) was 2193 mph (Mach 3.2+)…a world record, which still stands today as the fastest operational aircraft.
2. On 25 May 1961 President John F. Kennedy addressed Congress, telling them…"I believe this nation should commit itself to achieving the goal, before this decade is out, of landing a man on the moon and returning him to Earth safely." The conventional wisdom asked why and said, we are not ready. The leadership turned to American ingenuity, and Neil Armstrong (shown in Fig. 11.2) walked on the moon on 20 July 1969 and returned to earth safely.

Fig. 11.1    SR-71.

Fig. 11.2    Neil Armstrong on the moon.

Fig. 11.3   F-117A.

3. The F-117 Nighthawk at 66 ft in length, a wing span of 43 ft and a TOGW of 52,000 lb has the same RCS value as a ball bearing. Fifty-nine were built for the USAF for tactical strike. The aircraft first flew in June 1981, became operational in October 1983, and was retired in April 2008. The F-117A is shown in Fig. 11.3.

# Chapter 12

## EPILOGUE

Nowhere do the Lessons Learned talk about faster and bigger computers, more and better analysis codes or paneling, and lofting tools. The Lessons Learned have nothing to do with the mechanics of doing the conceptual design…they influence the mental state of the design team's minds. The computer and the codes let the design team do a better and more thorough analysis, deeper design trades, and higher-fidelity CFD and structural design. But putting a top-of-the-line computer system on a weak design approach is like taking a pig to the beauty parlor. When you are finished, it is still a pig.

If the Lessons Learned discussed in this book are implemented with a passion, the results should be a superior winning design approach…an eagle. You can still take the eagle to the beauty parlor…but it doesn't need it.

One final point before we close. There is no danger that the computer can ever replace the human in the design loop. The Lessons Learned that we discussed cannot be programmed into a computer as a set of instructions. The computer cannot decide who the customer is, then listen to that customer and ferret out the MoM, then challenge the requirements, and finally negotiate with the customer a credible set of requirements. These interactions are absolutely critical in producing a winning design approach and are the purview of the human alone. So sleep well tonight…your job is safe.

---

#### IT AIN'T OVER 'TIL THE FAT LADY SINGS

People ask me why there isn't an 11th Lessons Learned. I reply that I don't really have an answer…but if I had an 11th Lessons Learned it would be the title of this factoid. I could not close out this book without telling the story of the slickest piece of entrepreneurship that I encountered during my career. It had to do with how General Dynamics, Convair Division in San Diego became the producer of the AGM-129 Advanced Cruise Missile (ACM).

DARPA started the black ACM program (code word TEAL DAWN) in 1976 with a selected source MS 0 conceptual design study, The contractors in the conceptual design study were Boeing, Northrop, and

McDonnell. General Dynamics, Convair Division was invited to participate but declined saying they were too busy with the development of the Tomahawk. Boeing won the down select in mid-1977 and was awarded a three-year $36M MS 1 validation contract. I had just finished the GBU-15 Analysis of Alternatives in AF/Studies and Analysis and was asked to join DARPA as the TEAL DAWN Program Manager. I had strongly recommended to DARPA that they carry two contractors into the validation phase as competition was a good thing. DARPA agreed, but money was tight and so we proceeded with a single contractor – Boeing.

In mid-1978 I got a visit request from Jim Beggs, General Manager of GD/Convair, to discuss a proposition. Jim Beggs said that their decision not to participate in TEAL DAWN was a poor one and so he offered the following proposition:

GD/Convair would agree to do the same tasks, schedule, and deliverables (an MS 2 validated design ready for FSD) as Boeing for $6M. Convair would spend $30M of company funds (5 to 1 funding split) and provide DARPA with competition and access to the Tomahawk experience. Beggs realized that his company was a year behind Boeing in a three-year race, but his people were anxious for the challenge.

I ran the proposition through DARPA security and legal. We offered the same deal to Northrop and McDonnell, but they were not interested. We accepted Convair's proposal and overnight we had a competitive program. I watched the activity in the Boeing and Convair camps. The passion and zeal at Convair was impressive and they made up a one-year late start in four months. Boeing felt confident, after all they had a one-year head start. The MS 2 FSD down select was in mid-1980 with Convair selected as the winner.

TEAL DAWN transitioned to the USAF in 1982 as the AGM-129. Convair went on to produce 460 AGM-129As for the USAF.

So don't ever count yourself out of anything. There is always a way to get back into the race. Convair put together an innovative but risky plan, Jim Beggs sold it to DARPA and the Air Force...the fat lady sang...and Convair went to Disney Land.

# REFERENCES

[1] "Acquisition of Major Defense Systems," DoD Directive 5000.1, U.S. Department of Defense, U.S. Government Printing Office, Washington, DC, 2009.
[2] "Code of Ethics For Engineers," Accreditation Board for Engineering and Technology (ABET), Washington, DC, 5 Oct. 1977.
[3] "Code of Ethics for Engineers," National Society of Professional Engineers (NSPE), Alexandria, VA, July 2007.
[4] Carichner, G. E., and Nicolai, L. M., *Fundamentals of Aircraft and Airship Design: Volume II Airship Design and Case Studies*, AIAA, Reston, VA, 2013.
[5] "Long Fall for Pentagon Star," *Washington Post*, Washington, DC, 13 Nov. 2004.
[6] "$615,000,000...Cheap at the Price," *Aviation Week and Space Technology,* New York, 22 May 2006, p. 66.
[7] Bond, D., "Druyun Aftershocks," *Aviation Week and Space Technology*, New York, 9 May 2005, p. 19.
[8] Mullin, S. N., "The Evolution of the F-22 Advanced Tactical Fighter," 1992 Wright Brothers Lecture, AIAA Paper 92-4188, AIAA, Reston, VA, 1992.
[9] Kaminski, P., et al., *Pre-Milestone A and Early-Phase Systems Engineering*, National Academics Press, Washington, DC, 2008, p. 20055.
[10] Nicolai, L. M., and Carichner, G.E., *Fundamentals of Aircraft and Airship Design: Volume I Aircraft Design*, AIAA, Reston, VA, 2010.
[11] Richey, G. K., *F-111 Systems Engineering Case Study*, Center for System Engineering, Wright-Patterson AFB, OH, 10 March 2005.
[12] Olds, R., *Fighter Pilot*, St. Martin's Press, New York, 2010, p. 10010.
[13] R&M Proof, "Avionics Gear Installed Under Rear Seat in F-4," *Aerospace Daily*, Vol. 135, No. 15, 23 Sept. 1985, p. 113.
[14] Fulghum, D., "Roche on the Warpath," *Aviation Week and Space Technology*, 3 March 2003, p. 56.